World Class Applications of Six Sigma

T0361219

World Class Applications of Six Sigma

Edited by

Jiju Antony
Ricardo Bañuelas
Ashok Kumar

Routledge
Taylor & Francis Group

LONDON AND NEW YORK

First Published by Butterworth-Heinemann
This edition published 2011 by Routledge
2 Park Square, Milton Park, Abingdon, Oxfordshire OX14 4RN
711 Third Avenue, New York, NY 10017
Routledge is an imprint of the Taylor & Francis Group, an informa business

First edition 2006

Notice
No responsibility is assumed by the publisher for any injury and/or damage to persons
or property as a matter of products liability, negligence or otherwise, or from any use or
operation of any methods, products, instructions or ideas contained in the material herein.
Because of rapid advances in the medical sciences, in particular, independent verification
of diagnoses and drug dosages should be made

British Library Cataloguing in Publication Data
A catalogue record for this book is available from the British Library

Library of Congress Cataloguing in Publication Data
A catalogue record for this book is available from the Library of Congress

ISBN 13: 978 0 7506 6459 2
ISBN 10: 0 7506 6459 2

Typeset by Charon Tec Ltd, Chennai, India
www.charontec.com

This book is dedicated to Frenie and Evelyn, Clara, and Manjusha

Contents

List of editors and contributors

Editors

Jiju Antony is a Professor at the Centre for Research in Six Sigma and Process Improvement (CRISSPI), Glasgow Caledonian University, UK.

Ricardo Bañuelas is a Logistics Improvement Leader and Black Belt at Rolls-Royce plc, UK.

Ashok Kumar is a Professor in the Department of Management, Seidman School of Business, Grand Valley State University, Michigan, USA.

Contributors

Alan Harrison is the Head of Business Improvement at Weir Pumps, Glasgow, Scotland.

Alex A. Balbontin is the Vice President and Master Black Belt, JP Morgan Chase, UK.

Arun Raychaudhuri is the Head of ASPIRE Project Competency Group, Tata Steel, Jamshedpur, India.

Asim Choudhary is a Manager of Coke Plant at Tata Steel, Jamshedpur, India.

David Bigio is an Associate Professor of Mechanical Engineering at A.J. Clark School of Engineering, University of Maryland, USA.

D.P. Deshpande is the Chief of Project, Battery No. 10 and By-product Plant, Tata Steel, Jamshedpur, India.

Edgardo Escalante is a Professor in the Department of Industrial Engineering at the ITESM, Monterrey, Mexico.

Edith Ng is a Research Assistant based in the Department of Industrial Engineering and Engineering Management at Hong Kong University of Science and Technology, Hong Kong.

Fugee Tsung is an Associate Professor in the Department of Industrial Engineering and Engineering Management at Hong Kong University of Science and Technology, Hong Kong.

Jaideep Motwani is a Professor and Chair in the Department of Management, Seidman School of Business, Grand Valley State University, Michigan, USA.

K.Y. Lam is with the Logistics Cargo Supervisors Association, Hong Kong.

Lisa Lopez is an Executive Director of Community Services at Commonwealth Health Corporation, Kentucky, USA.

Maneesh Kumar is a Doctoral Student at the Centre for Research in Six Sigma and Process Improvement (CRISSPI), Glasgow Caledonian University, UK.

Manoj Tiwari is a Professor at the Department of Forge Technology, NIFFT, Ranchi, India.

Martin Brace is a Master Black Belt within one of the world's leading diversified technology companies.

Nick Shubotham is a Design Team Leader and Certified Black Belt within one of Europe's leading domestic appliance manufacturers.

Ricardo Días Pérez is a Quality Manager in a multinational chemical company in Mexico.

Richard So is an Associate Professor based in the Department of Industrial Engineering and Engineering Management at Hong Kong University of Science and Technology, Hong Kong.

Rick Edgeman is a Professor and Chair in the Department of Statistics, University of Idaho, USA.

Ronald D. Snee is a Principal with Tunnell Consulting, PA, USA.

Suman Biswas is the Chief of ASPIRE, Tata Steel, Jamshedpur, India.

Sung H. Park is a Professor in the Department of Statistics, Seoul National University, Korea.

Supratim Halder is a Manager of Coke Plant at Tata Steel, Jamshedpur, India.

T.S. Li is a Safety Officer based in the Safety and Environmental Protection Office, Hong Kong University of Science and Technology, Hong Kong.

Thomas Ferleman is the President of Ferleman and Associates, Management and Technology Consulting, Maryland, USA.

Acknowledgments

As editors and as chapter authors, we have benefited from the advice and help of a number of people in the preparation of this edited book. This collection of ideas on Six Sigma case studies was conceived during the year 2003–2004 when Jiju had finished his third book on *Design of Experiments for Engineers and Scientists*. When he took his ideas to Ricardo and Ashok, they foresaw the potential which resulted in the present volume. Jiju's work on this book reflects his experiences and lessons learned from his previous books *Experimental Quality and Understanding, Managing and Implementing Quality* and the one mentioned above.

When it comes to people, unfortunately no list can ever be complete and someone will be omitted. We hope those we do not mention specifically here will forgive us. We are intellectually indebted to the many academics and practitioners whose research and writing have blazed new trails and advanced the discipline of Six Sigma. We are most grateful to the reviewers who made valuable suggestions that guided our preparation of this edited book.

It is our sincere hope that by reading this book you will find something new or at least think about some things in a new way. As always we welcome your thoughts about this book. Your suggestions, comments and feedback regarding the coverage and contents will be taken to heart, and we will always be grateful for the time you take to call our attention to printing errors, deficiencies and other short shortcomings.

Co-authors, Ricardo and Ashok not only contributed case studies to this book but also toiled relentlessly to assure that what we have learned through good and bad experiences are reflected in the book. Our Publisher Elsevier had the patience of Job in helping get this book to market. For all of the many people with Elsevier who have helped us – a big thank you.

Finally, the editors would like to acknowledge the following publishers for having given permission to reproduce case studies, figures and tables in the book:

IMechE – Journal of Engineering Manufacture – Professional Engineering Publishing
International Journal of Six Sigma and Competitive Advantage – Interscience Publishers
Six Sigma Forum Magazine – American Society for Quality (ASQ)
Quality and Reliability Engineering International Journal – John Wiley and Sons.

Professor Jiju Antony

Introduction

There are extensive numbers of textbooks and articles available on Six Sigma. This book is certainly not another textbook on Six Sigma, but rather a book which demonstrates how Six Sigma could be applied or has been applied in a real industrial setting. This book will be written on the assumption that the reader is already acquainted with the subject area and therefore the philosophical and theoretical aspects of Six Sigma will be limited in this book. This book is essentially a collection of case studies taken from multinational manufacturing and service corporations, and therefore offers a more specialist treatment of various aspects of Six Sigma which are either not covered or only briefly covered in existing books available in the market. What we are offering here is a book which will allow the reader to appreciate some of the complexities and problems associated with the implementation of Six Sigma methodology, some of the key tools and techniques of Six Sigma in contemporary organizations.

This book also provides a systematic approach using the Six Sigma methodology (Define–Measure–Analyze–Improve–Control or DMAIC) for both understanding and assessing the application of various tools and techniques (why, how, etc.) of Six Sigma. The purpose of this book is to clearly demonstrate how the Six Sigma methodology can be applied to a particular manufacturing or service or transactional problem, and also the calculations of estimated cost savings from the project. This book is primarily aimed at advanced undergraduates, postgraduates/post-experience students, quality management and improvement practitioners, Six Sigma practitioners and researchers engaged in Six Sigma.

Before we provide an executive summary of the main issues arising from the chapters, we felt that it was important to give an executive introduction to Six Sigma covering aspects such as definitions, what makes Six Sigma different from other quality improvement initiatives, benefits of Six Sigma, the Six Sigma methodology (DMAIC) and the future of Six Sigma. We will encourage readers to refer to other Six Sigma textbooks for more detailed information on the theoretical implications of Six Sigma.

What is Six Sigma?

Six Sigma is widely recognized as a business strategy that employs statistical and non-statistical tools and techniques, change management tools, project management skills, teamwork skills and a powerful roadmap (DMAIC) to maximize an organization's return on investment (ROI) through the elimination of defects in processes. We explain the term 'Six Sigma' in both statistical and business terms. In statistical terms, *Six Sigma* implies 3.4 defects or mistakes or errors or failures per million opportunities. Here *Sigma* is a term used to represent the variation about the average of a process. The focus of 'Six Sigma' is not on counting the defects in processes, but the *number of opportunities* within a process that could result in defects. For instance, consider a call center or contact center and for any given call center from a customer to the contact center, the following opportunities might lead to defects, which ultimately causes customer dissatisfaction and hence lost customers:

- The manner in which the customer is greeted by the customer service agent or customer service representative.
- The accuracy of information provided by the agent to the customer.
- The queuing time before the customer gets hold of an available agent.
- The number of rings before an agent responds to the call.
- The accuracy of the data entry of customer identity to retrieve past data.
- The listening, speaking and interpretive skills of the agent.
- The accuracy of data entry if a fault or problem has been reported by the customer.
- The time taken to restore the service if a fault has been reported.
- The manner in which the call is ended.
- The timely arrival of any requested follow-up material, etc.

The objective of a Six Sigma strategy in the above case is to understand the process within the call center which creates the defects and devise process improvement methods to reduce the occurrence of such defects which improve the overall customer experience. The focus must be on four issues:

- What is the nature of the defects which are occurring in the process?
- Why are such defects occurring and at what frequency?
- What is the impact of a defect on customers?
- How can these defects be measured and what strategies should be implemented to prevent the occurrence of such defects?

In business terms, Six Sigma is defined as 'a business strategy used to improve business profitability, to drive out waste, to reduce costs of poor quality and to improve the effectiveness and efficiency of all operations so as to meet or even exceed customers' needs and expectations.' The Six Sigma approach begins with a business strategy and ends with top-down implementation, having a significant impact on profit if successfully deployed. It takes users away from *intuition-based decisions* (what we think is wrong) to *fact-based decisions* (what we know is wrong).

It is important to note that a Six Sigma quality level of performance should not be the goal for all processes. A lower Sigma quality level of performance may be acceptable for some processes. For example, a credit card company had a target that 95% of customers wishing to speak to an available customer service agent or representative must be connected within six rings. The company had established through a customer survey that customers were willing to wait up to seven or eight rings provided they were informed by a recorded voice that a customer service agent or representative would attend to their queries soon. The company also found through research that a further reduction to five or less rings would not increase customer satisfaction significantly. In such cases, we do not really need a Six Sigma process capability. On the other hand, in some processes, even Six Sigma may not be enough. For example, the Sigma quality level for an airline industry for safe landing could be higher than Six Sigma.

The average performance of most processes today is in the range of 3–4 Sigma quality level. The Six Sigma measure of process capability assumes that the process mean may shift over the long term by as much as 1.5 Sigma, despite our best efforts to control it. In the Six Sigma process, 3.4 defects per million opportunities (DPMO) are obtained by assuming that the specification limits are six standard deviations away from the process target value and that the process may shift by as much as 1.5 Sigma. The 3.4 DPMO value is the area under the normal curve beyond $6 - 1.5 = 4.5$ Sigma. Similarly, the 66807 DPMO for the 3 Sigma process is the area under the normal curve beyond $3 - 1.5 = 1.5$ Sigma.

What makes Six Sigma different from other quality management/improvement initiatives?

We personally have seen that senior management in many organizations view Six Sigma as another quality improvement initiative or flavor of the month in their list. We are often told by many engineers and managers in small and big companies that there is nothing really new in Six Sigma compared to other quality initiatives we have witnessed in the past. In response, we often ask a simple question to people in organizations who practice total quality management (TQM), 'what do you understand by the term TQM?' We often get many varying answers to this question. However, if we ask a bunch of Six Sigma practitioners, 'what do you know of the term Six Sigma?' we often get an answer which means more or less the same thing that we would have expected. In our opinion, the following aspects of the Six Sigma business strategy are not accentuated in previous quality improvement initiatives:

- Six Sigma strategy places a clear focus on achieving measurable and quantifiable financial returns to the bottom-line of an organization.
- Six Sigma strategy places an unprecedented importance on strong and passionate leadership, and the support required for its successful deployment.
- Six Sigma methodology of problem-solving integrates the human elements (culture change, customer focus, belt system infrastructure, etc.) and process elements (process management, statistical analysis of process data, measurement system analysis, etc.) of improvement.
- Six Sigma methodology utilizes the tools and techniques for fixing problems in business processes in a sequential and disciplined fashion. Each tool and technique within the Six Sigma methodology has a role to play and when, where, why and how these tools or techniques should be applied is the difference between success and failure of a Six Sigma project.

- Six Sigma creates an infrastructure of Champions, Master Black Belts (MBBs), Black Belts (BBs) and Green Belts that lead, deploy and implement the approach.
- Six Sigma emphasizes the importance of data and decision-making based on facts and data rather than assumptions and hunches!
- Six Sigma utilizes the concept of statistical thinking and encourages the application of well-proven statistical tools and techniques for defect reduction through process variability reduction methods (e.g. statistical process control and design of experiments).

Benefits of Six Sigma

Organizations adopting Six Sigma business strategy will have the following benefits:

- Effective management decisions due to heavy reliance on data and facts instead of gut-feelings and hunches. Hence costs associated with fire-fighting and misdirected problem-solving efforts with no structured or disciplined methodology could be significantly reduced.
- Increased understanding of customer needs and expectations, especially the critical-to-quality (CTQ) service performance characteristics which will have the greatest impact on customer satisfaction and loyalty.
- Increased cash flow by making processes more efficient and reliable.
- Improved knowledge across the organization on various tools and techniques for problem-solving, leading to greater job satisfaction for employees.
- Reduced number of non-value-added operations through systematic elimination, leading to faster delivery of service, faster lead time to production, faster cycle time to process critical performance characteristics to customers and stakeholders, etc.
- Reduced variability in process performance, product capability and reliability, service delivery and performance, leading to more predictable and consistent level of product quality and service performance.
- Transformation of organizational culture from being reactive to proactive thinking or mindset.
- Created new customer opportunities, improved market position relative to competitors, etc.
- Improved internal communication between departments, groups, etc.
- Improved cross-functional teamwork across the entire organization, employee morale and team spirit.

Six Sigma Methodology

The five-stage methodology of Six Sigma begins with the Define phase. The *Define* phase involves identifying a project's CTQ characteristics driven by the voice of the customer (VOC), followed by developing a team charter and finally defining a high-level process map connecting the customer to the process and identifying the key inputs and requirements.

In the *Measurement* phase, the team identifies the key internal processes that influence CTQ characteristics and measures the defects currently generated relative to those processes. The project's CTQ characteristics are selected with the help of fishbone diagrams, quality function deployment (QFD), Pareto charts, etc. A gauge repeatability and reproducibility (GRR) study must be carried out to ensure that the measurement system is capable and acceptable. In service-related processes, GRR for attribute data is an option.

The *Analysis* phase consists of three steps (fundamentally). However, it heavily depends on the process and the type of business you are involved in. The first step is to establish process capability which measures the ability of the process to meet customers' specifications. The second step is to define performance objectives by the team benchmarking and identifying the sources of variation by performing analysis of variance (ANOVA) tests and hypothesis testing. Based on the above information, the root causes of defects and their impact on the business/process can be identified.

The *Improvement* phase helps the team to confirm the key variables or causes and quantifies their effects on the CTQs. In this phase, one may develop potential solutions to fix the problems and prevent them from recurring. Once the potential solutions are developed by the team, it is advisable to evaluate the impact of each potential solution using a criteria–decision matrix. Solutions that have a high impact on customer satisfaction and bottom-line savings to the organization need to be examined to determine how much time, effort and capital will need to be expended for implementation. It is also important to assess the risk associated with each potential solution. Techniques such as design of experiments, Taguchi methods, response surface methods, SERVQUAL, etc. are useful in this phase.

The *Control* phase consists of the following steps:

- Develop corrective actions to sustain the improved level of process performance (product/service/transactional).
- Develop new standards and procedures to ensure long-term gains.
- Implement process control plans and determine the capability of the process.
- Identify a process owner and establish his/her role.
- Verify benefits, cost savings/avoidance.
- Document the new methods.
- Close project, finalize documentation and share the key lessons learned from the project.
- Publish the results internally (monthly bulletins) or externally (conferences or journals) and recognize the contribution made by the team members.

A perspective on the future of Six Sigma

Six Sigma has been sweeping the business world with remarkable results to the bottom-line of many organizations since its adoption in the late 1980s, driving both continuous and breakthrough improvements in process performance and product/service quality. Based on the evidence of several textbooks, white papers, journal articles, workshops, conferences and the amount of training going on, Six Sigma still appears to be on the way up. We personally feel that Six Sigma will evolve over time like many other initiatives we have seen in the past. However, the key concepts, the principles of statistical thinking, tools and techniques of Six Sigma will stay for many years, irrespective of whatever the 'next big thing' will be. I also believe that the existing Six Sigma toolkit will be enriched by the continuous emergence of new useful tools and techniques, especially in the software, finance and healthcare applications. Some of the emerging trends of Six Sigma include: integration of Six Sigma and lean thinking, integration of Six Sigma with EFQM Excellence Model, Theory of Constraints, ISO 9001: 2000, Six Sigma for small- and medium-sized enterprises (SMEs), Six Sigma for the public sector, Six Sigma and its impact on organizational performance, leadership and its impact on Six Sigma success, etc.

Many companies in our opinion are still tackling the low-hanging fruits using the Six Sigma DMAIC methodology. This is the case with many service-oriented companies (healthcare, financial services, etc.) and SMEs. Six Sigma does not appear to have yet peaked on a global basis, in that it is still growing in many European and Asian countries, particularly in countries such as UK, Germany, Sweden, India, China and Middle East. We have also observed that very few universities in Europe are engaged in teaching and research on Six Sigma. This needs to be changed in the future so that collaborative Six Sigma research projects between the academic and industrial world must be established in both engineering and business schools. The academic world has a crucial role to play to bridge the gap between the theory and practice of Six Sigma, and to improve the existing methodology of Six Sigma.

Six Sigma will keep on to build its momentum in almost all types of industries, irrespective of the size and turnover, with no signs of giving up in the immediate future. The challenge for all organizations is to integrate Six Sigma into their core business processes and operations rather than managing it as a separate initiative. Six Sigma might evolve into a 'new package' when it fails to deliver significant impact to the bottom-line of organizations. However, the sound principles and key concepts of Six Sigma will stay with it for many years. We also perceive that the existing cognitive gap in the statistical knowledge required by the engineering fraternity can be easily bridged by introducing Six Sigma as a core module to their curriculum. As a final remark, we believe that organizations developing and implementing Six Sigma strategy should not view it as an advertising banner for promotional services.

Part I of this book is focused on the applications of Six Sigma in manufacturing sector. This part consists of seven chapters. Chapter 1 provides a comprehensive review of Six Sigma implementation at Dow Chemical Company. The authors have taken a careful and exhaustive look at Dow's Six Sigma program – from the time the idea of a cultural change germinated in the minds of Dow leadership that led to Six Sigma implantation till today when the program has been fully implemented and the results are in. Chapter 2 presents an application of Six Sigma to reduce waste in a coating process. This chapter describes in detail how the project was selected, how the methodology was applied, and how various tools and techniques of Six Sigma have been employed to achieve substantial financial benefits. Chapter 3 provides an interesting Six Sigma case study with one of the leading steel companies in India. The application of Six Sigma methodology has resulted in a breakthrough improvement to the company. A significant improvement in Sigma quality level was achieved for one of the critical processes which led to immense reduction in defect rate for the process under study. Chapter 4 illustrates the systematic application of Six Sigma DMAIC methodology in one of the largest European white goods producers. The problems are usually manifested in the field when the product fails to perform according to the design intend under specific conditions for the desired period of time, causing customer dissatisfaction, increasing service and warranty costs and decreasing sales. The savings generated from this project were estimated to be over US $500,000 annually. Chapter 5 deals with a case study performed in an automobile company to reduce the defect rate in casting product/process using the DMAIC problem-solving methodology. The application of the Six Sigma methodology reduced the number of defects in the casting process and thereby improved customer satisfaction and business profitability. The estimated savings generated from this project were around US $111,000 per year. Chapter 6 focuses on the application of the Six Sigma methodology in a company that manufactures coal products, with the general objective of improving a specific process and therefore to satisfy its customers' requirements. The specific objective of this study was to analyze the elements that constitute the Six Sigma methodology, and to apply it to improve a specific coal manufacture process where the highest amount of scrap is generated.

Part II of this book is focused on the applications of Six Sigma in service sector. This part consists of five chapters: Chapters 7 to 11. Chapter 7 demonstrates the application of Six Sigma in a healthcare setting. This case study is focused on the use of DMAIC methodology on surgical site infections to reduce the infection rate and thereby achieving increased patient satisfaction. Chapter 8 illustrates the Six Sigma program within Doosan Heavy Industries & Construction Company based in Korea. This chapter discusses the Six Sigma curriculum within the company, Six Sigma project selection process, application of DMAIC, etc. This chapter also presents some of the similarities and fundamental differences between Lean and Six Sigma methodologies. Chapter 9 introduces a detailed example of the use of Six Sigma within a leading global financial services firm. This chapter highlights the project challenges with regard to data collection and soft issues. The author of this chapter finally discusses the challenges of applying Six Sigma in banking and the future trends of the adoption of this approach in this industry sector. Chapter 10 illustrates the application of Six Sigma to reduce the number of fall incidents among cargo handlers working on top of cargo containers in a cargo handling industry in Hong Kong. The application of the DMAIC methodology provided a rigorous approach to the company in order to meet increasingly high safety standards and an eventual near-zero hazard rate. Chapter 11 presents an interesting case study on Six

Sigma applied to information technology (IT) system. The estimated long-term savings from this project is between 2 and 3 million dollars.

Part III of this book is focused on the applications of Six Sigma in transactional environments. This part consists of two chapters: Chapters 12 and 13. Chapter 12 provides a detailed application of Six Sigma to increase the accuracy of newspaper. This case study resulted in the errors being reduced by 65%, producing a savings in time at the copy desk of more than $226,000 per year. This was a cost avoidance rather than a hard dollar savings as the persons involved were freed up for other more value-added work and did not leave the employment rolls. The actual savings were even more because of the effects of fewer errors in the composing room, fewer pages needing to be redone and fewer press stops. Chapter 13 describes application of Six Sigma DMAIC approach in human resources (HR) function and processes within an engineering organization. Six Sigma in HR function has demonstrated values and benefits of proactive behavior, customer focus and alignment.

Part I

Applications of Six Sigma in Manufacturing Sector

1

Six Sigma implementation at Dow Chemical Company: a comprehensive review

Ashok Kumar, Jaideep Motwani and Jiju Antony

> *A combination of a mindset, a set of methodologies, and a tool set which*
> *is positioned to accelerate the implementation of business strategies*
>
> Dow Chemical Company

1.1 Introduction

The quality of a company's products and processes has an intrinsic relationship with its financial and strategic performance. A US General Accounting Office study of the 20 highest-scoring *Malcolm Baldrige National Quality Award* applicants suggested that these organizations achieved better quality, lower cost, greater customer satisfaction, improved market share, and higher profitability compared to their competitors (GAO, 1991). Using a database of 3000 companies representing over 16,000 years of data, the Profit Impact of Market Strategy study (W-13)[1] concludes that, 'In the long run, the most important single factor affecting a business unit's performance is the quality of its products and services relative to those of its competitors.'

Yet, quality remained the Cinderella of the competitive weapons available to American and West European businesses up until the mid-1980s. Indeed, the contemporaneous operations strategy literature is replete with strategic formulations which focused solely on a single variable, overly simplistic dynamic of competition (lower-the-price, higher-the-profits) to obtain competitive advantage. Unfortunately, this paradigm failed in the face of newer paradigms which called for strategy formulations that drew from a richer and more exhaustive set of variables, such as quality, flexibility, and agility (Fine and Hax, 1985; Swamidass, 1986; Skinner, 1996a,b) to gain higher market share, and hence, higher profitability.

---

[1] The references for web sites follow a specific format in this chapter. W-x represents reference number x in the web site references placed at the end of this chapter.

Many corporate (Fine and Hax, 1985; Skinner, 1996a,b) strategists ascribe the erosion of American leadership of auto and electronic industries in the 1980s to this failure of strategic vision, especially the ignorance of the potential of quality as a formidable competitive weapon between the 1960s and 1980s. Conversely, the Japanese companies who actually demonstrated a greater grasp of changing strategic paradigms, gained market share in the automobile industry from 4% in the 1960s to 36% in the 1980s, much to the chagrin of American manufacturers.

It was in this context that American businesses began to perforce expand their strategic vision and examine the quality of their products, processes, practices, and programs in earnest. In 1984, Motorola, a cell phone manufacturer, embarked on a Six Sigma journey under the leadership of Michael Harry, who set forth the goal of improving the quality of products in the Government Electronic Group (GEG) from Three Sigma (66,800 defects per million opportunities) level to Six Sigma level (3.4 defects per million opportunities). Breaking away from traditional methodologies of problem-solving that have shown, quite often, limited success, Harry devised his own methodology that is known as DMAIC (Define–Measure–Analyze–Implement–Control). While DMAIC works best for existing product and process improvements, Design For Six Sigma (DFSS) works best when an entirely new product or process has to be designed or a new concept operationalized. Motorola's Six Sigma program was characterized by the following key ingredients:

(a) a primary goal of total customer satisfaction;
(b) common uniform quality metrics for all areas of the business;
(c) identical improvement rates for all areas of the business measured on one scale;
(d) goal-directed incentive for managers and employees;
(e) coordinated training in reasons for these goals and ways to achieve them (Kumar and Gupta, 1993).

They defined the Six Sigma quality system as collective plans, activities, and events designed to ensure that the products, processes, and services satisfy customer needs. In summary, Six Sigma is a customer-focused approach to business that provides an overall framework for quality management.

Since its inception at Motorola, Six Sigma has been adopted by many companies with significant financial gains. The following chart (Table 1.1) provides the year in which the Six Sigma program began in certain companies and Table 1.2 summarizes the financial gains from Six Sigma implementation at certain well-known companies (W-11) (see footnote 1 on p. 3).

Due to Six Sigma's rigorous problem-solving prowess, the program is well suited to bring about targeted improvements in all strategic priorities and not just cost and quality. Indeed, in customer-focused analyses, Six Sigma has registered a record of commendable results for the companies listed in Table 1.2, and many others such as Sony, ABB, Texas Instruments, Citicorp, Chase Manhattan, Caterpillar, Raytheon, and Bombardier Transportation. Well-executed Six Sigma projects have consistently delivered sizeable improvements in customer-related performance characteristics: critical-to-quality (CTQ), critical-to-cost (CTC),

Table 1.1 Year of inception of Six Sigma at pioneering companies

Company name	Year of Six Sigma inception
Motorola	1986
AlliedSignal (merged with Honeywell in 1999)	1994
GE (NYSE: GE)	1995
Honeywell (NYSE: HON)	1998
Ford (NYSE: F)	2000
Dow Chemical Company	1999

Table 1.2 Savings from well-known Six Sigma programs

Year	Revenue ($ billion)	Savings ($ billion)	Revenue savings (%)
Motorola			
1986–2001	356.9	16.1	4.5
AlliedSignal			
1998	15.1	0.52	9.9
GE			
1996	79.2	0.2	0.2
1997	90.8	1	1.1
1998	100.5	1.3	1.2
1999	111.6	2	1.8
1996–1999	382.1	4.43	1.2
Honeywell			
1998	23.6	0.5	2.2
1999	23.7	0.6	2.5
2000	25	0.7	2.6
1998–2000	72.3	1.84	2.4
Ford			
2000–2002	43.9	1.6	2.3
Dow Chemical Company			
1999–2002	120	1.5	1.25

critical-to-delivery (CTD), and critical-to-responsiveness (CTR) (W-2). For this reason, Six Sigma is no longer just a quality program; it is considered a powerful strategic tool that can deliver performance improvements in all areas of strategic priorities (cost, quality, flexibility, and delivery). Due to its pervasive application potential, it has also become a philosophy, an approach in all the undertakings of a company rather than just a tool or technique for specific applications. R. Balu of Fast Company explains why Six Sigma has emerged as a methodology for driving business strategy (W-3). According to her, the following three critical components of Six Sigma philosophy constitute essential characteristics of Strategic Six Sigma Initiative (SSSI):

- *DFSS* generates new processes, products, and/or plants.
- *DMAIC* improves existing products and processes.
- *Process management* enables leverage and sustains gains achieved.

In this chapter, we describe the implementation of Six Sigma at Dow. We begin at the very beginning when Dow had started searching for a catalytic program to accelerate its quest for improvement in its strategic position and attainment of core corporate values. We discuss the preparatory work done to ensure smooth launching of Six Sigma through the Staircase model. We specifically identify the unique features of Dow's Six Sigma, such as its singularly dedicated customer focus, leverage, and customer loyalty. We then provide summary of approaches and results of several Six Sigma case studies conducted by Dow. The studies show the breadth of scope of Six Sigma implementation; it encompasses all aspects of the business functions and processes at Dow (manufacturing, service, and staff) and the impressive contributions resulting from these studies reflect the strategic success of Dow's Six Sigma effort.

1.2 Dow Chemical Company

The Dow Chemical Company is one of the world's largest science and technology companies. It serves nearly 70,000 customers worldwide in 180 countries, supplying 3,200 distinct products to a host of domestic and international markets, including the food, transportation, health and medicine, personal homecare, and building and construction markets. With an annual sale of approximately $33 billion, Dow is currently organized into 15 major businesses and is engaged in more than 40 joint ventures. It employs approximately 50,000 employees in 208 manufacturing sites in 38 countries.

Dow produces and markets an impressive range of products: pharmaceuticals, automotive, adhesives and sealants, agricultural, building and construction, imaging/photography, paints and coatings, paper, plastic and rubber products, textiles, chemical processing, electronics and telecommunication, textiles, and medical. Dealing in such a wide range of products, markets, cultures, and regions is truly a challenging task that requires a *culture* of doing things right and improving on the *status quo* constantly.

Dow's mission is to 'constantly improve what is essential to human progress by mastering science and technology.' Given the mission, commitment to the *triple bottom line* of economic prosperity, environmental stewardship, and corporate social responsibility, and a vast scope of products and markets, Dow is in perpetual need of a result-oriented program that can effectively respond in real time to its dynamically evolving customer needs and competitors' capabilities. Fueled by such a need, and after virtually year-long deliberations, Dow embarked on its Six Sigma journey in 1999.

1.3 The Six Sigma program at Dow

1.3.1 Preamble and preparation

Process change typically begins with strategic initiatives (often included in the corporate strategic plan) from the senior management team (Kotter, 1995). These could be a reaction to a need (e.g., company's inability to provide adequate customer service) or a proactive push to leverage potential opportunities (Earl, 1994). Evidence also exists that strategic change, and arguably process change, is often incremental, informal, emergent, and based on learning through small gains (Mintzberg and Waters, 1985) versus being revolutionary and radical.

From its very inception in 1897, Dow has designed, developed, and implemented programs that serve to enhance its competitive position through enhancement of its core values. In the early 1990s, Dow instituted value-based management tools whose primary goal was to analyze its business processes from the standpoint of the value they create for its customers and the resources expended in creating such values. The application of value-based tools served to improve Dow's products and processes to maximize customer satisfaction while minimizing the cost. During the same period, Dow also employed quality performance mechanisms and business process re-engineering to accomplish its strategic goals of enhancing its core values.

In 1994, the company reframed its business strategy, known as *Strategic Blueprint*. This blueprint consisted of four interrelated components: competitive standards, value growth, culture, and productivity. The blueprint provided unprecedented strategic advantage to Dow, bestowing it with the global leadership in chemical industries. The leadership set its eyes on achieving global leadership in all areas of their business.

1.3.2 The Six Sigma journey

The beginning

The Strategic Blueprint, which reshaped Dow's strategy in 1992, paid off handsomely in terms of Dow's rapid movement forward to gain the world leadership in the chemical industry and making significant inroads in other sectors such as food, medical, transportation, and personal homecare. The significant milestones implemented subsequently over the next 6 years helped a great deal in Dow's pursuit of excellence in improving performance, productivity, and value, but the pace was not good enough for Dow's leadership. In 1998, they began to explore for an *enabler* program that would deliver the breakthrough they were looking for. The search culminated in a program that had delivered salutary results for companies like Motorola and General Electric (GE) who were in the similar boat as Dow, at least in terms of productivity, quality, and people's issues. The breakthrough program of great promise was none other than Six Sigma!

Implementation planning

The implementation planning for Six Sigma at Dow began with a 4-month planning period to make sure a well-thoughtout strategy of implementation could be put in place that impacts the *people* of Dow and brings about the positive cultural change that is needed for higher levels of performance, productivity, and values. It was decided that Six Sigma at Dow would not be a corporate-level program to be pushed down the throats of the business units with a lot of responsibility and very little authority. It was conjectured at the leadership level that for best results the business units would integrate Six Sigma in their respective business strategies, thus placing accountability for success or failure squarely on the shoulders of the company's unit leaders.

Following a series of meetings with Six Sigma Academy in early 1999, the implementation of Six Sigma began at Dow in earnest. Initial projects were chosen carefully for maximum impact and, true to expectations, these projects delivered stupendous results in terms of Dow's core Six Sigma objectives: cost, quality, and customer satisfaction. Motivated by the results, Dow decided, late in the summer of 1999, to expand the implementation across the board (i.e., to all its business units across the world). Kathleen Bader was named to the post of Executive Vice-President for Quality and Business Excellence, and was charged with leading the effort.

1.3.3 Dow's Six Sigma methodology at variance: loyalty, leverage, and Seven Sigma

Bader developed a model for implementation of Six Sigma; the lowest stair of this staircase was the *vision* and the top, success. There were, in all, 10 stairs (see Figure 1.1). In this section, we elaborate on two extensions of the basic DMAIC terminology that are unique to Dow: loyalty and leverage.

Loyalty

In Dow's model of Six Sigma implementation, D (define) phase of the DMAIC methodology is replaced by L (loyalty), since it was believed that the Six Sigma program must conform to and promote the central Dow value of striving for total customer satisfaction. In turn, this would imply that all Six Sigma projects would be motivated primarily by the objective of enhancing customer loyalty through improved customer satisfaction. A key strategy to obtain greater customer loyalty was what Bader called *Outside-in-Focus*. This meant

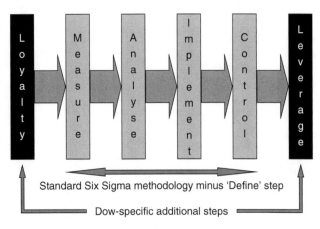

Figure 1.1 Dow's Six Sigma methodology.

that Six Sigma projects would have to be grounded in the information collected both from outside (especially markets) and as the customers (outside) looked at the company (inside).

Leverage

Leverage is defined as the effective multiple implementation of demonstrated best practices. Thus, a solution to a specific problem developed in a specific project would be communicated through their central communication system to all of the Dow business units, which, in turn, could benefit from that solution wherever appropriate, thus saving precious resources expended in *reinventing the wheel*. This *multiplier effect* is called leverage by Dow. To appreciate the power of leveraging, Jeff Schatzer, 'Dow's former Six Sigma Director of Communication is worth quoting: ...' Dow has 20 polystyrene plants, 4 polycarbonate plants, 10 acrylonitrile butadiene styrene (ABS) plants, 27 polyethylene plants, 4 polyethylene terephthalate (PET) plants, 2 polypropylene plants, 2 toluene diisocyanate (TDI) plants, and 1 co-polymer plant; that is, 70 plants that are engaged in production of plastics products. Add to this 15 blending facilities, contract manufacturers, and joint ventures around the world and it is safe to say that Dow dots the globe. Over the past several years, Dow has established global work processes, global techN centers, and global business structures. Our unique information system, the global platform, allows us to communicate the best practices anywhere in the world at the speed of a key stroke. ... The 'multiplier effect of leveraging Six Sigma through our global MET (most effective technology) process is delivering powerful results.'

Seven Sigma

For Dow, Six Sigma is not good enough when it comes to environmental health and safety. Dow aims at performing at Seven Sigma level for those processes which involve ergonomic injury or sickness. Seven Sigma translates into 20 defects per billion opportunities.

1.3.4 The 'Staircase of Change Leadership' model for Six Sigma implementation

To bring about the changes needed for implementation of Six Sigma, Dow used the Staircase of Change Leadership model. This model consists of 10 steps as shown in Figure 1.2.

We will discuss these staircase steps in the context of Six Sigma implementation at Dow.

Vision

Dow's vision of its Six Sigma program was projected in its 1999 Annual Report. It is: 'Dow will become recognized and lauded as one of the premier companies of the 21st century, driven by an insatiable desire to achieve a Six Sigma

Figure 1.2 Staircase of Change Leadership.

level of performance and excellence in all we do.' In the same report, Dow also projected an earning before interest and tax (EBIT) of $1.5 billion cumulatively over the course of the next 4 years.

Values

We have earlier discussed Dow's values in terms of triple bottom line. In this model, Dow defines the following as its corporate values:

- Integrity
- Respect for people
- Unity
- Outside-in-Focus
- Agility
- Innovation

Interestingly, Six Sigma platform directly supports each of these values. It is also noteworthy that the last two corporate values of Dow directly tie in with the major competitive priorities – flexibility and delivery – on which Dow competes in the 21st century.

Attitude

According to Bader, 'Visionary leadership is rarely accidental. It is the attitude that imposes accountability, inaugurates change, inspires belief, invokes commitment, and induces results.' To that extent, Six Sigma is as effective as the mindset of the people implementing it, and attitude assumes an enormous importance as a variable for Six Sigma success.

Language

According to Bader, 'the soul of attitude is evidenced in language.' There are two aspects of language. The aspect that leadership used to facilitate commissioning

Six Sigma as a program of change was, by design, *solution-oriented, positive* language. The language that involved developing project solutions for chosen Six Sigma projects was institutionalized in consonance with the standard Six Sigma language.

Behaviors

Having prepared the attitude and determined the language, change agents (includes all employees – leadership as well as lower-level personnel) must behave in accordance with the objectives of the change. Bader provided the following key elements of behavior as guidelines for Six Sigma implementation:

- Intolerance for variation
- Measuring both inputs and outputs to ensure high levels of efficiency as well as performance
- Accountability for all
- Delivering measurable, sustainable gains in terms of Dow's values
- Delivering customer satisfaction to build loyalty
- Leveraging competitive advantage through information sharing

Best practices

Well-known success stories of Six Sigma implementation were studied carefully to identify what worked and what did not. In addition, Dow carried out its own detailed analysis of what would work best for Dow to make its Six Sigma program outstandingly successful. This yielded two factors, loyalty and leveraging, which we have addressed in previous sections. All success factors including the two 'L's were incorporated in Six Sigma training programs.

Articulated strategy   In order to bring about institutional changes in Dow's culture and values at the fastest pace, a well-articulated strategy was needed. Six Sigma would bring such changes in good time, but drivers of change needed Dow's cultural metamorphosis with greater urgency. The Six Sigma breakthrough strategy was designed to respond to this urgency. This strategy consisted of systematically following three sequential business processes: stages of change, management of change, and management of implementation.

Change process begins by unfreezing the existing (pre-programmed) behavior, followed by instilling new behavior, institutionalizing new behavior, and finally refreezing the institutionalized behavior into a sustained phase (W-10). Drawing on several organizational change studies, Dow determined that the following aspects of Six Sigma would need to be managed to make sure that the transformation would occur in the right direction and magnitude:

- Vision, values, and strategy
- Processes and measures of outcomes
- Organizational culture

- Information technology (IT) and systems
- Human resource policies

Finally, managing implementation required advanced strategic planning which was flexible enough to respond to evolving changes in the organizational aspects listed above, while at the same time setting up long-term objectives consistent with the desired rate of transformation. Dow decided to use the *Hoshin Kanri* approach to develop a matrix of objectives that could be monitored on an annual basis. *Hoshin Kanri*, for the benefit of interested readers, is a systems approach to the management of change in critical business processes using a step-by-step planning, implementation, and review process. *Hoshin Kanri* provides a planning structure that will bring selected critical business processes up to the desired level of performance.

1.3.5 Six Sigma project organizational structure

Dow's hierarchical structure for each Six Sigma project is no different from that of most other leading companies. The following are the typical project management hierarchy, roles, and responsibilities associated with any Six Sigma project at Dow (W-7):

- *Sponsor*: Business executive leading the organization.
- *Champion*: Responsible for Six Sigma strategy, deployment, and vision.
- *Process Owner*: Owner of the process, product, or service being improved – responsible for long-term sustainable gains.
- *Master Black Belts*: Coach Black Belts, expert in all statistical tools.
- *Black Belts*: Work on three to five $250,000 per year projects; create $1 million per year in value.
- *Green Belts*: Work with Black Belts on projects.

1.4 Six Sigma tools, techniques, software, and IT used at Dow

1.4.1 Tools and techniques used by Dow for Six Sigma projects

For Dow, Six Sigma is '… a combination of a mindset, a set of methodologies, and a tool set which is positioned to accelerate the implementation of business strategies.' Also, Six Sigma is a fact-based, data-driven process of problem-solving. As such, certain tools and techniques are recommended that the Six Sigma teams use to reach the solution. Table 1.3 provides an exhaustive list of tools and techniques that are selectively used by Dow's Six Sigma. However, Dow's approach is not limited to any given set of tools or techniques. Their Six Sigma mindset virtually calls for unlimited tools and techniques … whatever it takes to solve the problem. We provide a case study later in this chapter that uses simulation to solve a logistical problem arising from a high anticipated demand at one of Dow's business units.

Table 1.3 Tools and techniques used for Six Sigma program

Technical	Non-technical
Statistical	
• Gauge R&R studies	• Scatter plots
• Statistical process control	• Histograms
• Design of experiments	• Box plots
• Response surface methods	• Run charts
• Regression analysis	• Multi-vari charts
• Process capability analysis	• Tally sheets
• Robust design + tolerancing	• Normal probability plots
• Sampling methods	• Correlation studies
Non-statistical	
• Quality function deployment	• Process flowcharting
• Quality costing	• Project management skills
• Root cause analysis	• Team chartering
• Process + design FMEA	• Leadership skills
• Benchmarking	• Brainstorming
• Mistake proofing	• People issues
• Pareto analysis	
• Affinity diagram	

1.4.2 Six Sigma software

Many project teams at Dow use JMP software for their analysis although it is not a requirement. According to Obermiller, 'Dow were looking for a statistical software that would meet the needs of 95% of the people 95% of the time. The product needed to work under Windows and be affordable.' Furthermore, it was stipulated that the software should have the capability of conducting data analyses using typical as well as advanced statistical techniques and should have ease of use, accuracy, documentation, expandability, customizability, and price. JMP was selected out of 17 potential candidates.

1.4.3 IT and Six Sigma at Dow (W-10)

IT is perhaps the most important enabler of a business process change, especially for a company that is as large and diverse as Dow. Dow has used IT effectively since the 1960s. In the 1960s, it was used for streamlining and automating chemical plants, in the 1980s for streamlining and optimizing operations in business units in chemical plants and beyond, and in the 1990s for optimizing business processes globally across all of the Dow business units. Today, it is a factor of paramount importance in making Six Sigma a huge success, especially with its leveraging role for multiplier effect. It has brought about positive and synergistic effects for Dow by converting it from fragmented pieces to a unified whole by increasing standardization by using a single technical information architecture for the entire organization, and by providing a unique leveraging capability across the corporation. The interplay between IT and Six Sigma at Dow is so strong that Mike Costa, Dow's Global Director, Six Sigma Work Processes and Expertise Center, says that the DFSS methodology is *baked into*

Dow's IT development methodology. What does it do for Dow's Six Sigma program? It enables important activities listed below required for DFSS methodology to be done effectively in an optimized manner:

- Project definition through well-defined multi-generation plans (MGPs); MGPs are long-term project implementation plans that have a time line for completion of important steps of the project;
- Project management through voice-of-customers (VOC) and quality function deployment (QFD); these are two very important strategic tools because they enable leadership to listen to customers and thus improve customer satisfaction;
- Helps with alternative analyses of projects by facilitating use of such tools as Pugh matrix or criteria-based matrix; the Pugh matrix is a tool used to facilitate a disciplined, team-based process for concept generation and selection. Several concepts are evaluated according to their strengths and weaknesses against a reference concept called *the datum* (base concept). The datum is the best current concept at each iteration of the matrix. The Pugh matrix allows you to:
  - compare different concepts
  - create strong alternative concepts from weaker concepts.
- Arrive at an optimum concept that may be a hybrid or variant of the best of other concepts.

Figure 1.3 shows how IT is integrated into DMAIC methodology to create new IT solutions.

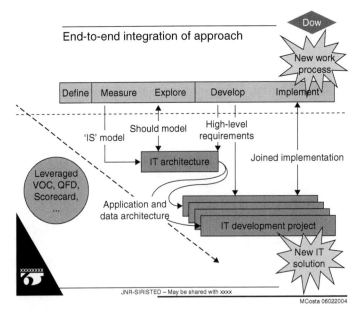

Figure 1.3 Developing new IT solutions at Dow (the Six Sigma way). *Source*: Mike Costa, June 2004.

The capability maturity model

Dow uses the capability maturity model (CMM) to evaluate its IT performance. It is currently at level 3, called the *defined* level; that is, its IT service processes are documented, standardized, and integrated into standard service processes. All services are delivered using approved, tailored versions of the organization's standard service processes (W-12). It is now moving toward level 4, which is called the *managed* level.

1.5 Where is Dow on its Six Sigma journey?

Dow does not share its Six Sigma performance data generally, particularly not the aggregate data. The following facts were, however, unearthed through web search and personal conversation with certain high-level officials related to Six Sigma at Dow as of Summer 2003:

- Global full-scale implementation beginning in 2000 that extended to virtually every business, site, and area
- 300 master Black Belts, 1400 full-time Black Belts, and 2500 Green Belts
- 41.7% employees engaged in successful Six Sigma projects
- Nearly 3000 projects completed (gains realized)
- More than 4000 active projects are in progress
- Average estimated project gains = $600,000
- Average project completion time = 6 months
- The goal of achieving EBIT of $1.5 billion through Six Sigma poised for accomplishment almost a year ahead of time in 2003.

1.6 Selected case studies

As previously stated, Dow's implementation of Six Sigma could justifiably be called *all-encompassing* for at least two reasons. First, the objectives and scopes of Six Sigma projects extended well beyond the limited quality improvement objective. Indeed, the program derived its mission from the broadest goal of maximizing customer satisfaction at the expense of the least amount of resources. Translated into operational parlance, this implied that Six Sigma projects set forth their objectives not only as cost minimization or quantum improvements in quality; they also concentrated on such strategic aspects as improving customization and organizational agility. Second, the scope of projects was unlimited and cut across all functional boundaries. Thus, Six Sigma projects were applied to manufacturing, service, and staff functions uniformly within and across the boundaries of such functions. As a result, Six Sigma became a program that impacted every fiber and the entire culture of the company. More than a tool or technique, it became a philosophy that pervaded the entire fabric of the organization in terms of being a driver for perfection. We now provide a synoptic view of several representative studies that Dow undertook along its Six Sigma journey with spectacular results. The depth and breadth of these studies evidences the exhaustiveness of the scope of Six Sigma application at Dow.

1.6.1 Energy conservation studies (DMAIC approach) (W-4a,b)

Steam trap improvement project

Define:	Optimize steam delivery in energy systems by reducing steam loss through steam traps.
Measure:	The 75 failed traps and 45 visible steam leaks identified.
Analyze:	Prioritize steam traps and pressure applications based on annualized steam loss.
	Leverage (Dow uses this term in the sense of intelligently using information/solutions developed in other projects) information on inverted bucket traps.
	Validate that 600-lb steam is superheated.
Implement:	After repair on culprit traps and with a preventive maintenance program in place, steam loss reduced by an estimated 87.3%.
Control:	Develop a predictive preventive maintenance schedule for traps so that steam loss can be monitored and problems fixed in time.
Bottom-line impact (estimated):	$200,000 per year.

Polycarbonate unit energy reduction project

Define:	Freeport polycarbonate plant in Texas consumed twice the amount of energy per pound of product in Dow's German plant.
Measure:	The 17% increase in energy usage from 2000 to 2001.
Analyze:	Identified root causes as steam, nitrogen, air, and electricity losses.
Implement:	Reduced energy usage by 10%. Repaired sources of leaks. Designed campaign for change of mindset to reduce building electricity usage.
Control:	Maintenance of devolatilazation system monitor identified leaks. Promote energy-waste reduction mindset.
Bottom-line impact (estimated):	$500,000 per year.

System unit energy envelopment project

Define:	Identify opportunities to optimize plant energy heat integration, improve efficiency, and reduce emissions of carbon monoxide, carbon dioxide, and nitrogen oxides.
Measure:	Average boiler efficiency of 69%. 143,000 lb per hour of wasted condensate.

Analyze:	Factor analysis to quantify major main effects. Pinch study to identify heat integration opportunities. Engineering and statistical tools to analyze and optimize processes.
Implement:	Improve distillation efficiency to obtain 17 MM BTU per hour. Improve process furnaces efficiency to obtain conversion energy reduction by 20 MM BTU. Improve boiler efficiency to obtain energy usage reduction by 25 MM BTU per hour. Improved heat integration to reduce energy usage by 18 MM BTU per hour.
Control:	Ongoing.

1.7 An ergonomic case study: Minimizing musculo-skeletal disorders (DMAIC approach) (W-6a,b)

As previously indicated, Dow's implementation of Six Sigma encompasses many and varied areas of application. This particular study is about how Dow's design and construction (DDC) business unit, which is responsible for managing the design and construction of Dow's facilities worldwide, used the Six Sigma approach to bring about a 90% reduction in the company's reportable injury and illness rate (i.e., to a level of 0.24). DDC employs about 1,250 employees who spend the bulk of their time working in front of computer terminals. The study is significant because, at average, each injury causes 9 days of productivity loss and the yearly losses from such injuries ranges in scores of millions of dollars. DDC management wanted to redundant to prevent absenteeism from injury and illness before it grew out of hand. According to Dow's 2003 Annual Report, they improved their injury and illness rate by 19% in 2003 and 78% since 1994, which is better than the target they had set forth. Furthermore, 60% of their plants had no injuries at all in 2003. Following is a brief account of the DMAIC steps of the study that began in 2000:

Define:	The losses due to injury and illnesses resulting from poor ergonomics could range in millions of dollars. The define phase presented unique difficulties because at the time of this study, the automotive division already had a very low rate of injuries (three in 1999). It was decided to define the problem in terms of causal variables and bring about an improvement in the environment that causes injuries/illness.
Measure:	Three sets of key variables that influence injury and illness rate were identified as: user attributes (e.g., amount of time at terminals), user behavior (e.g., posture, force, duration of use), and environmental factors. A 70% improvement in all ergonomics-related factors that lead to injury/illness was targeted.
Analyze:	Data was collected on work environment, workstation design, user behavior, and user training. Using several Six

Sigma analytical tools (e.g., Ishikawa diagram, work performance matrix, antecedent–behavior–consequence analysis, balance of consequence analysis), a list of root causes was developed. Using Pareto and other Six Sigma tools, the list of probable causes was narrowed down and validated. A partial list of the underlying causes is as follows:

- Employees not connecting work design (ergonomics) and their health and safety;
- Lack of adjustability of furniture for customized use;
- DDC does not emphasize ergonomics enough.

Improve: The following improvements were carried out:

- Workstations upgrade-plans were developed to deal with the customization problems.
- Workstation ergonomics were improved through increasing employee accountability.
- Awareness was also increased by posting personal testimonials and stories of other employees who suffered injury due to ergonomic problems.
- At each facility, specially trained *volunteer* employees were designated as ergonomic focal points who would receive work-design-related complaints and also voluntarily identify opportunities for ergonomic improvements in work design.
- Safety first mentality was stressed and publicized.

Control: Each of the above improvements was subjected to regular monitoring and followup. Ergonomic focal points were used to carry out personalized followup with other employees.

1.7.1 *Key benefits from Six Sigma project*

The study resulted in the following measurable improvements:

- Severity of ergonomic injury has significantly declined. In 2001, 53% of injuries resulted in lost work time. In 2003, only 30% injuries were quite as severe.
- A 64% reduction in risk factors was observed since the base line measurement was done.
- Dow is well on the way to reach the target injury and illness rate of 0.24 by 2005.
- The results were gainfully leveraged throughout the Dow Chemical Company for:
  - reduction of repetitive stress injuries,
  - reduction of motor vehicle accidents,
  - improved safety for visitors,
  - site logistics risk reduction,
  - off-the-job safety process improvement.

1.8 Lessons and insights from Dow's Six Sigma program

According to the literature, successful application of Six Sigma involves planning, effort, and flexibility. Simply applying another company's plan, problem-solving process, team structure, or training package does not ensure success. For Dow, implementation of Six Sigma was replete with lessons learned. Here are a few gems that Dow picked up from its experience of implementing a highly successful Six Sigma program.

1.8.1 Constancy of purpose

There must be a *constancy* of purpose that should be the governing principle for bringing about desired change through Six Sigma implementation. The entire leadership should be committed to this purpose. The purpose itself must evolve from the core values that the implementing organization wants to espouse. Dow laid down its core values and the direction of change that it wanted to bring about to all its leaders. Mike Parker, the CEO of Dow, made sure that he used every public opportunity to display his sense of total commitment to Six Sigma.

1.8.2 Financial rigor

It is important that the results of Six Sigma are rigorously monitored, analyzed, evaluated, and validated. Dow instituted business rules and established a team of trained financial analysts to review and validate financial benefits from its Six Sigma projects. Applying financial rigor to project financial offers transparency and credibility to the company's implementation of Six Sigma.

1.8.3 Data capture and knowledge management

A successful Six Sigma implementation needs a flexible, user-friendly, and integrated-across-the-enterprise database. Six Sigma drives a data-based decision-making process. Dow has invested significantly in the construction and maintenance of its database system for Six Sigma. Data should also be used and applied through a centralized communication system to maximize the leveraging opportunity.

A way to do work ... not additive

Six Sigma is not an extraneous add-on program that can be superimposed on other programs. In order to be successful, it must be *the way in which work is done*. Indeed, in its ultimate form it is a mindset that drives the way people identify and solve problems. On a collective basis, it becomes a philosophy that engages everyone in a synergistic manner to contribute to the core values of the organization and becomes a source of sustained competitive advantage.

Pipeline conundrum

Keeping a robust pipeline is essential to maintaining and building momentum for Six Sigma implementation. Reflecting back on its decision to implement via rapid transformation, Dow would consistently say that the approach it took was the right one at the right time. Time spent up front in creating a project pipeline would be well spent.

Start small

It is best to start implementation in small pockets that have great potential for improvement. The results from small, limited implementation become Six Sigma's best ambassadors to the rest of the organization.

Training

The importance of adequate and focused (e.g., role-related) training for successful implementation cannot be overemphasized. Training expenses should be viewed as perhaps the best investment opportunity because the knowledge management of the information pertaining to Six Sigma projects could be seriously compromised without proper training and attitude building. Dow set forth a goal of training at least 3% of 70,000 employees – an idea that has paid handsome dividends over the years.

1.9 Concluding remarks

In this chapter, we have taken a careful and exhaustive look at Dow's Six Sigma program, from the time the idea of a cultural change germinated in the minds of Dow leadership that led to Six Sigma implantation until today, when the program has been fully implemented and the results are in. Simply based on the results on Six Sigma-related EBIT, wherein they set forth a goal of $1.5 billion in 1999 and accomplished it roughly a year ahead of time, one would have to say that the program has been quite successful. In terms of the metrics of Six Sigma program growth, it is noteworthy that the Six Sigma-related personnel grew in impressive numbers. Based on the information in the public domain with regard to number of Black Belts, Green Belts, champions, and above all, the number of employees who contribute to Six Sigma projects has registered a very impressive growth. There is no question that the commitment from the top (one has to look at Dow's last CEO, Mike Parker's public pronouncements and exhortations of Six Sigma in every speech) to make Six Sigma a mindset and a work culture has been a goal well accomplished. A grassroots program like Six Sigma needs a tremendous amount of preparation prior to launching. The strategic planning done by Kathleen Bader in preparing Dow leadership and lower-level employees is very well done indeed. All one has to do is look at her Staircase model to see how such planning should be done in a rather short period of time. It is also noteworthy that Dow has allowed its Six Sigma program to flourish across the board – its applications are as diverse as Dow itself; the applications have been implemented in service and staff functions, which is refreshing and indicative of Dow's capacity to realize Six Sigma's full potential. In a similar vein, it must be stated that Dow has used Six Sigma not only 'for dollars', but also to improve other laudable goals such as environment, health, and safety. This is certainly commendable and makes their Six Sigma program strategic as opposed to simply tactically or operationally effective.

Is everything perfect then in regards to Six Sigma implementation? Not quite. First and foremost, Dow has about 41.9% employees involved in Six Sigma projects and they have set a goal to involve everyone by 2005. It is possible that this goal will be achieved, but not quite certain, given the situation at this point. More importantly, there is very little information available in the public domain about how well the Six Sigma program at Dow is delivering its other value goals. For instance, we do not know the status of Dow's customer satisfaction. Was there a metric created that computed Dow's customer satisfaction as a composite score? the Jury is still out on that account. All Six Sigma programs must emanate from and promote the core values of an organization. There is anecdotal evidence that Dow has implemented their program to achieve that goal. Yet, there is no clear system-wide metric that is publicly reported to substantiate this conjecture. All in all, a great program with great results!

Acknowledgments

The authors would like to thank Mr. Jeff Schatzer, Six Sigma Communication Leader at Dow, for the information he provided in his PowerPoint presentations, and in his informal communications with the authors and Grand Valley State University Master's degree students.

References

Earl, M. (1994). Viewpoint: new and old business process redesign. *Journal of Strategic Information Systems*, 3(1), 5–22.

Fine, C.H. and Hax, A.C. (1985). Manufacturing strategy: a methodology and an illustration. *Interfaces*, 15(6), 28–46.

GAO (1991). *Management Practices: U.S. Companies Improve Performance Through Quality Efforts*, Washington, DC, General Accounting Office, GAO/NSIAD, 91-190.

Kotter, J. (1995). Leading change: why transformation efforts fail. *Harvard Business Review*, 73(2), 59–67.

Kumar, S. and Gupta, Y. (1993). Statistical process control at Motorola's Austin assembly plant. *Interface*, March–April, 23, 84–92.

Mintzberg, H. and Waters, J. (1985). Of strategies deliberate and emergent. *Strategic Management Journal*, 6, 257–272.

Skinner, W. (1996a). Manufacturing strategy on the "S" curve. *Production and Operations Management*, 5(1), 3–14.

Skinner, W. (1996b). Three yards and a cloud of dust: industrial management at century end. *Production and Operations Management*, 5(1), 15–24.
Swamidass, P.M. (1986). Manufacturing strategy: its assessment and practice. *Journal of Operations Management*, August, 471–484.

Worldwide web references

1. Taleo Case Study. http://www.taleo.com/en/knowledge/media/pdf/CaseStudy_Dow.pdf
2. Antony, J. Customer Centered Six Sigma Initiatives. http://www.qualityamerica.com/knowledgecente/articles/Antony_CUSTOMERCENTRICSIXSIGMA.pdf
3. Balu, R., Strategic Six Sigma, Fast Company, May 2001. http://media.wiley.com/product_data/excerpt/47/04712329/0471232947.pdf
4. Tannenbaum, K., Dow Chemical Company and The Office of Energy Efficiency and Renewable Energy of the US Department of Energy, Applying Six Sigma Methodology to Energy-Saving Projects.
   a. http://www.oit.doe.gov/showcasetexas/pdfs/casestudies/cs_dow_sixsigma.pdf
   b. http://www.eere.energy.gov
5. Delivering Quality Products and Services Using JMP for Six Sigma.
   a. http://www.qualitydigest.com/june03/departments/01_app.shtml
   b. http://www.jmp.com information on JMP
   c. http://www.jmp.com/news/applications/jmpapp_dow.pdf
6. Ergonomics Case Study, Dow Chemical Company's Six Sigma Case Study.
   a. http://www.osha.gov/SLTC/ergonomics/dow_casestudy.html
   b. http://www.eere.energy.gov
7. Quality and Environmental: Dow Automotive. http://www.dow.com/automotive/quality
8. Six Sigma for IT Service Management, White paper by Proxima Technology, August 2003. http://www.proxima-tech.com/products/whitepapers/Six_Sigma_for_SLM.pdf
9. Buss, P. and Ivey, N., Dow Chemical Design for Six Sigma Rail Delivery Project, Proceedings of the 2001 Winter Simulation Conference, B.A. Peters, J.S. Smith, D.J. Medeiros, and M.W. Rohrer (eds). http://ieeexplore.ieee.org/xpl/tocresult.jsp?isNumber=21078
10. Costa, M., Capturing IT Value through Leveraging Six Sigma and Other Strategic Approaches, June 2004. http://www.isssp.com/media/LC04/preentation_slides
11. Financial Performance of Six Sigma Companies.
    a. Motorola Six Sigma Services. Motorola University, July 22, 2002. http://mu.motorola.com/sigmasplash.htm
    b. GE Investor Relations Annual Reports. General Electric Company, July 22, 2002. http://www.ge.com/company/investor/annreports.htm
    c. Honeywell Annual Reports. Honeywell Inc., July 22, 2002. http://investor.honeywell.com/ireye/ir_site.zhtml?ticker=HON&script=700
    d. Better Understand Six Sigma Plus With Honeywell's Special PowerPoint Presentation. Honeywell Inc., 22 July 2002. http://www.honeywell.com/sixsigma/page4_1.html
    e. *Quality Digest*, 'Six Sigma at Ford Revisited', June 2003, p. 30. http://www.qualitydigest.com/june03/articles/02_article.shtml
    f. AlliedSignal Inc. 1998 Annual Report, Honeywell Inc., July 22, 2002. http://www.honeywell.com/investor/otherpdfs/ALD98.pdf

12. The IT Service Capability Maturity Model. http://www.itservicecmm.org
13. Production Impact of Market Strategy. http://www.pims.com

List of Dow personnel referred to in this chapter

Schatzer, J.: Communication Leader for Six Sigma at Dow.

Bader, K.: Executive Vice-President for Quality and Business Excellence, Implementation Leader of Six Sigma at Dow.

Obermiller, D.: Master Blackbelt of Corporate R&D Six Sigma at Dow.

Costa, M.: Global Director, Six Sigma and Work Processes Expertise Center.

Parker, M.: CEO and Chairman (until 2003).

2

Manufacturing waste reduction using Six Sigma methodology

Ricardo Bañuelas, Jiju Antony and Martin Brace

2.1 Introduction

Six Sigma has been considered a powerful business strategy that employs a well-structured continuous improvement methodology to reduce process variability and drive out waste within the business processes using effective application of statistical tools and techniques. In this chapter the authors illustrate the use of DMAIC (Define–Measure–Analyze–Improve–Control) methodology in a step-by-step fashion to reduce waste in a film-coating process. It describes in detail how the project was selected, how the Six Sigma methodology was applied, and how various tools and techniques within the Six Sigma methodology have been employed to achieve substantial financial benefits.

Due to confidentiality agreement between the company and the authors, the name of the company cannot be revealed in this chapter. However, the company where the case study was performed is a large multi-national manufacturing corporation based in the UK. This chapter begins by describing how the project was identified and selected. After that, the chapter illustrates the different phases of the DMAIC methodology, and the tools and techniques employed to narrow the project and identify the sources of variation. With the successful application of the DMAIC methodology a team of Six Sigma practitioners saved more than US $120,000 (approximately £60,000) annually. The authors conclude this chapter by offering a brief summary and stating the lessons learned.

2.2 Project selection

Project selection is the process of evaluating individual projects or groups of projects, and then choosing to implement some set of them, so that the objectives of the organization will be achieved (Meredith and Mantel, 2000). Project selection is one of the most critical success factors influencing the outcome of Six Sigma projects. Selecting a project that is too large will cause valuable time

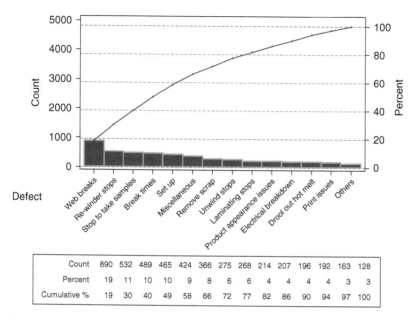

	Count	Percent	Cumulative %

Defect	Count	Percent	Cumulative %
Web breaks	890	19	19
Re-winder stops	532	11	30
Stop to take samples	489	10	40
Break times	465	10	49
Set up	424	9	58
Miscellaneous	366	8	66
Remove scrap	275	6	72
Unwind stops	268	6	77
Laminating stops	214	4	82
Product appearance issues	207	4	86
Electrical breakdown	196	4	90
Drool out hot melt	192	4	94
Print issues	163	3	97
Others	128	3	100

Figure 2.1 Pareto plot of defects.

to be lost during the define phase as Black Belts struggle to scope their projects and develop project charters that can be addressed using Six Sigma. In addition, projects should be linked to the right goals and impact at least one of the major stakeholders' issues. The company where this project took place has focused its Six Sigma effort in growth acceleration, cost reduction and cash flow improvement issues. In this manner, each of the company's business units has assigned different Six Sigma projects focused on the mentioned issues.

The primary objective of Six Sigma in this case is cost reduction through waste elimination/reduction. This effort is linked to the business' critical 'Y'[1] of reducing manufacturing costs. Having identified this opportunity the Six Sigma team, formed by one Black Belt and three Green Belts, focused on finding what was preventing one of the coating lines from achieving the business critical 'Y' goal. This coating line produces a wide range of coated products for the automotive market. The coating line is a continuous process with equipment designed to allow a non-stop production during roll changes of web materials and the unloading of finished rolls of production (Shephered, 1994). The Six Sigma team identified the number of line stops (stop offs) in the film-coating line and as an important indicator of the line performance. Figure 2.1 illustrates a Pareto plot for the number of stop offs assigned to different defects.

---

[1] The basic equation of Six Sigma, $Y = f(x)$, defines the relationship between a dependent variable (Y) or outcomes of a process and independent variables (the 'x's) or possible causes of problems associated with the process (Brue and Launsbry, 2003).

Gap analysis based on Q1 2002 data		
	Runtime gap (%)	Quality gap (%)
Run time/Gold star	**72**	**78**
Setups	8.13	2.9
Break times	5.31	1.8
Miscellaneous	5.17	3.7
Web breaks	3.98	5.0
Re-winder fails in chop-over	1.73	3.6
Appearance problems	1.71	1.4
Poor appearance	1.70	2.3
No print	0.67	1.6

Figure 2.2 Gap analysis.

Each line stop results in a subsequent line start, thus affecting the running time of the line. In addition, waste is generated during a line start as operational time is needed for the key process to reach steady state running. In order to determine the gap between what the film-coating line produces and what is required, a gap analysis was carried out. This analysis, shown in Figure 2.2, lists the potential causes preventing the achievement of the departmental critical 'Y.' The run time column states the different problems, which do not allow the line to run continually. The quality column represents the problems associated with quality issues.

The above causes prevent the coating line to reduce the waste and achieve the quality desired. Therefore, they are also seen as potential areas of improvement. However, DMAIC Six Sigma methodology is only recommended when the cause of the problem is unknown or unclear, the potential of significant savings exists and the project can be done in 4–6 months (Breyfogle *et al.*, 2001). In addition, it is important to prioritize potential areas of improvement using Six Sigma. To aim this, the team employed cause and effect matrix. This matrix, shown in Figure 2.3, lists all potential projects or opportunities that are likely to affect the outputs of the process, such as quality, waste and run time. In the cause and effect matrix outputs are listed in the top row and are assigned values according to the importance of customers and strategic business goals. In this case, they are quality, waste and run time. A high number indicates more importance. The first column contains the process step where the opportunity exists. At the intersection of each row and column, values are entered to quantify the amount of correlation, although to exist between the inputs 'X's and outputs 'Y's (9 being strong correlation and 0 no correlation). Numbers in each input are cross-multiplied with the importance number at the top of the column and summed across each row (Breyfogle, 1999). The opportunities or potential projects and their weightings in the matrix were based on the gap analysis, the Pareto plot and group expertize during brainstorming sessions.

As a result, a ranking for each opportunity provides the guidance that led the team to select three Green Belt projects. One of the projects has the objective

		Rating of importance to customer	6	3	1	
			Quality	Start up waste	Run time	Total
Process step	**Process opportunity**					
Coating	Flow problems		9	3	0	63
Coating	Raw material coater issues		9	0	0	54
Re-wind	Re-winder bump stop		3	9	3	48
Oven sprays	Web breaks		1	9	9	42
General	Poor appearance		6	0	1	37
Print press	No print		6	0	1	37
General	Poor brushing		3	1	3	24
General	Mechanical breakdown		3	0	0	18
Adhesive coating die	Adhesive supply problem		0	3	0	9
Autosplice	Creases		1	0	0	6
Print press	Print defects		1	0	0	6

Figure 2.3 Cause and effect matrix.

of finding and eliminating the causes of web breaks. Another is focused on improving product appearance. The third project is focused on identifying, quantifying and eliminating the source of variation that leads to failure due to spindle changes by the re-winder machine. The focus of this chapter is on the third project.

2.3 Six Sigma project: re-winder performance

The turret re-winder at the end of the coating line allows the line to run continuously. It winds up the 'web' of film in controlled tension producing large rolls of output. It frequently fails to change over from one roll to the other. The purpose of this Green Belt project is to identify, quantify and eliminate the source of variation that leads into failure to change over from one spindle or roll to another by the re-winder machine. The ultimate goal is to improve and sustain re-winder performance with well-executed control plans while reducing manufacturing costs.

As the project deals with the existing process, Six Sigma DMAIC problem-solving methodology was applied to the case. The project initially assumes that (Berryman, 2002; Huber, 2002; Nave, 2002):

- the design of the turret re-winder is essentially correct;
- customer or delivery partners needs are satisfied with that design;

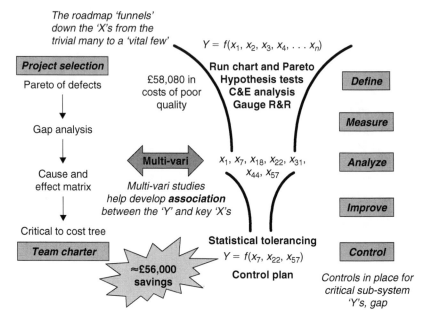

Figure 2.4 The roadmap.

- the current product/process configuration satisfies the functional requirements of the market and customer or delivery partner.

The following section describes how the five phases (DMAIC) of the Six
Sigma methodology have been applied to the problem at hand. These phases
are represented in the roadmap shown in Figure 2.4.

2.3.1 Define phase

This phase aims at defining the scope and goals of the improvement project in
terms of customer requirements and the process that delivers these requirements (Porter, 2001). Some of the deliveries of the define phase were incorporated during the project selection. By moving some of the define activity
forward into the project selection process the team was able to recognize the
company's goals, product-market domain and the basis of the competitive
advantage (George, 2002). It was also possible to breakdown high level 'Y's
into functional strategic areas by understanding the most critical cross-
functional activities of the company.

    The following four steps were performed during the define phase of the
DMAIC methodology:

1. Define the scope and boundaries of the project.
2. Define defects.

3. Define team charter.
4. Estimate the impact of the project in monetary terms.

Define the scope and boundaries of the project

This Green Belt DMAIC project links into a Black Belt project. Black Belt projects are generally boarder in scope than Green Belt projects. In this case, the Black Belt project focuses on the total coating line, whilst the Green Belt project is focused on one aspect, the turret re-winder performance.

Define defects

The defect is defined as failure of the re-winder to change from one spindle to the other. Each failure results in a loss of web tension and therefore a line stop.

Define project charter

Preparing a project charter requires team members to answer the following partially redundant elements; such redundancy in the elements helps team members distill the critical elements of the business case (Rasis *et al.*, 2002). The project charter was carried out to state the opportunity that exists. It summarizes the define stage from the business critical 'Y' and its linkage to the project 'Y.' It also cascades down the project description, goals and potential financial benefits.

Element 1: Process definition Element 1 aims at defining the process in which opportunity exists. This project is concerned about the performance of the re-winder chop-over process.

Element 2: Business critical 'Y' This element describes the opportunity as it relates to strategic business goal (Rasis *et al.*, 2002). Critical-to-quality (CTQ)/delivery/cost tree helps Six Sigma teams to move from general needs of the customers or business strategy to the more specific requirements (Eckes, 2001). Figure 2.5 depicts how generic business goals cascade into more specific potential Six Sigma projects. In this case, the critical business 'Y' is to reduce manufacturing costs. From this high level business 'Y' flows down the project 'Y,' which focuses on identifying, quantifying and eliminating the source of variation that leads to failure of the turret re-winder.

Element 3: Benefit impact Six Sigma projects should begin with the determination of customer requirements and it is essential to set project goals based on reducing the gap between the company's deliveries such as quality, delivery time, reliability and cost (Bañuelas and Antony, 2002). In this element, the anticipated impact of the project on overall business performance is estimated. Accordingly, the potential impact of the project is estimated based on three fundamental metrics: baseline performance, project goals and process entitlement. The Six Sigma project team decided that the re-winder should allow no

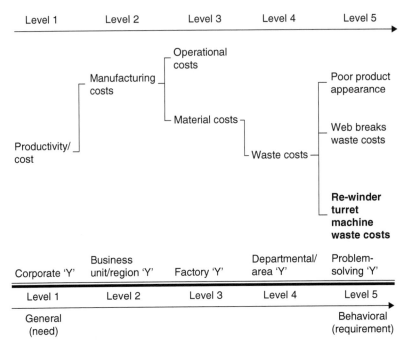

Figure 2.5 Critical to cost tree.

Table 2.1 Estimated financial benefits

	Baseline	Goal	Entitlement	Units
Stop offs	11.5	3	3	Occurrence/week or m²
% Gold star	62.5	75	75	%
Cost of poor quality	£1265	£330	£330	1 stop off = 80 m
				1 m waste = £110 waste
				material

Savings generated (estimated) = £60,000
Annual (based on 48 weeks)

more than its entitlement (i.e., three failures per week). As a result the calculated opportunities in financial terms are estimated from current performance (baseline) minus the targeted goals of the projects (see Table 2.1).

Element 4: Describe the scope and boundaries of the project   This project is limited to the re-winder and excludes line stops associated with other causes.

Element 5: List the key milestones activities with dates   The team decided to consider the DMAIC roadmap to list the key milestones of the projects.

Accordingly, the following phases were established with targeted dates to completion.

Element 6: Support required   In this element the team stated the support required from different departments. They highlighted the importance of obtaining maintenance expertize as a key to the success of this project.

Element 7: Core team members   The Six Sigma team in this case is a cross-functional one formed by a champion, a process owner, a Master Black Belt, a Black Belt, a Green Belt and six team members (people from maintenance, operators and a line supervisor).

As subsequent Six Sigma process steps of the DMAIC methodology are build on work completed during the define phase, the Six Sigma team ensured that the following deliveries have been achieved before proceeding to the next phase:

- Process linked to strategic business requirements.
- Customer and CTQ characteristics identified.
- Linkage of customer requirements to process outputs.
- Team formed with charter describing purpose, project plan, goals and benefits of the project.
- Financial benefits identified and calculated.

The Six Sigma team agreed to the listed deliveries and proceeded to the measure phase.

2.3.2 Measure phase

The measure phase has the purpose of mapping the current process and establish metrics that describe the project 'Y' in order to narrow the problem to its major factors or 'vital few' root causes (Pande *et al.*, 2000). The following steps were carried out during this phase:

1. Map process and identify process inputs and outputs.
2. Establish baseline process capability.
3. Establish measurement system capability.
4. Cause and effect analysis.
5. Data collection plan.

Map process and identify process inputs and outputs

Process mapping provides a picture of the steps that are needed to create the output or process 'Y.' It is a pictorial representation of the process, which helps to identify all value-added and non-value-added process steps, key process inputs ('X's) and outputs ('Y's) (Breyfogle, 1999). There are different tools suitable for process mapping, such as the flow chart, SIPOC (Supplier, Input, Process, Output, Customer) diagram and Standard Operation Procedures (SOP).

Issued by: R. Bañuelas	Process: Re-winder
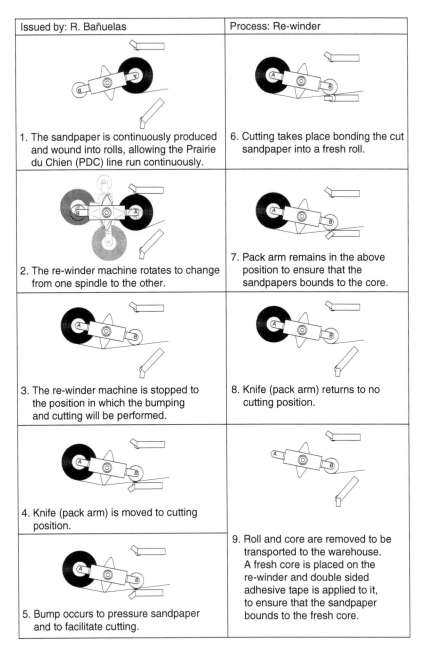 1. The sandpaper is continuously produced and wound into rolls, allowing the Prairie du Chien (PDC) line run continuously.	6. Cutting takes place bonding the cut sandpaper into a fresh roll.
2. The re-winder machine rotates to change from one spindle to the other.	7. Pack arm remains in the above position to ensure that the sandpapers bounds to the core.
3. The re-winder machine is stopped to the position in which the bumping and cutting will be performed.	8. Knife (pack arm) returns to no cutting position.
4. Knife (pack arm) is moved to cutting position.	9. Roll and core are removed to be transported to the warehouse. A fresh core is placed on the re-winder and double sided adhesive tape is applied to it, to ensure that the sandpaper bounds to the fresh core.
5. Bump occurs to pressure sandpaper and to facilitate cutting.	

Figure 2.6 Process mapping.

In this case a process map was generated using an SOP format. This format facilitates the understanding of the re-winder operation through a visual representation of the steps followed during this operation. The SOP was used throughout the course of the project as a reference (see Figure 2.6).

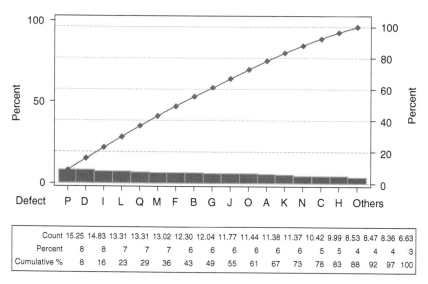

Count	15.25	14.83	13.31	13.31	13.02	12.30	12.04	11.77	11.44	11.38	11.37	10.42	9.99	8.53	8.47	8.36	6.63
Percent	8	8	7	7	7	6	6	6	6	6	6	5	5	4	4	4	3
Cumulative %	8	16	23	29	36	43	49	55	61	67	73	78	83	88	92	97	100

Figure 2.7 Re-winder Pareto.

Establish baseline process capability

A baseline indicates the current status of the process performance. The metric established for the re-winder's performance is a passed or failed chop-over (discrete data). From the manufacturing database the current process performance at a level of sigma 1.2 or 88.5% yield in the long term.

To record and display trends over the time and detect meaningful changes in the process, a run chart was created for the fraction of non-conforming chop-overs (Kiemele *et al.*, 1997). However, it showed no apparent ranking in failures over time. To recognize any possible trend in failures within the re-winder process, a Pareto plot of defects was generated (Kiemele *et al.*, 1997). Pareto plots create awareness for tackling the most pressing products first (Bicheno, 1994). It can be said from the Pareto plot in Figure 2.7 that most of the products have similar fraction of non-conformities. Therefore, it is difficult to assign any relationship between process and product performance.

Initially, the project 'Y' metric was discrete (i.e., pass/fail). The team strive to choose continuous data information over discrete information whenever possible, because continuous information provides more information for a given sample size (Breyfogle, 1999). However, the sample size required to deal with the fraction of non-conforming as continuous data can be large, especially for low non-conforming rates (Breyfogle, 1999). The cutting operation time of the re-winder seemed to be a good predictor of the operation outcome. The longer the operation cycle time, the higher the probability of failure. Conversely, when the cutting operation functioned properly, the operation cycle time appeared lower than that of the failures, as the box plot in Figure 2.8 illustrates.

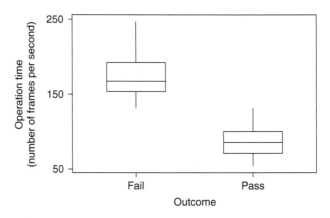

Figure 2.8 Box plot for operation time.

To be statistically confident that the cut cycle times were different for a successful chop-over vs. an unsuccessful chop-over, a hypothesis[2] test was carried out. In order to arrive at a valid conclusion a *t*-test was carried out with the following hypothesis:

$$H_0: \bar{x}_1 = \bar{x}_2 \quad \text{means of cutting operation time are the same.}$$

$$H_1: \bar{x}_1 \neq \bar{x}_2 \quad \text{means of cutting operation time are different.}$$

The hypothesis test was carried out at 5% significance level and 5% power. As the *p*-value was much lower than the significance level (5%), it was concluded that there was a significant difference in cut cycle time when the operation failed. From the hypothesis test, the team concluded that cut cycle time would be a good predictor of success in the winder operation. However, the measurement system needs to be analyzed to evaluate its potential capability.

Establish measurement system capability

Gauge repeatability and reproducibility (R&R) analysis was carried out to assess how much variation is associated with the measurement system (Kiemele *et al.*, 1997). According to gauge R&R analysis, the variation in measurement is subdivided into variation due to repeatability and variation due to reproducibility (Kiemele *et al.*, 1997; Pyzdek, 2001). The measurement system in this case has a sigma of 4.7892, which is formed by a sigma reproducibility of 0.8339 and a sigma repeatability of 4.8612. Thus, the repeatability is the main contributor to measurement system inconsistency. The sigma total is 27.907 producing a precision to total ratio of 0.1742 (4.81/27.907) that can be considered as marginal (Kiemele *et al.*, 1997). Therefore, the proposed measuring system shows a relative good measurement system capability.

---

[2] A statistical hypothesis is a statement about the parameters of one or more populations (Montgomery and Runger, 2003).

Cause and effect analysis

Having mapped the process, the team proceeded to analyze the potential causes of failure. Although this task is carried out in detail during the analysis phase, the team decided to start to discern the causes of the problem in order to identify 'X's to measure. Cause and effect analysis was carried out to illustrate the various causes that affect the re-winder performance (Ishikawa, 1974). Figure 2.9 illustrates the potential causes that could generate a failure in the re-winder cutting operation. They were identified in a team brainstorming session in which manufacturing engineers, maintenance team, line operators and supervisors participated. Figure 2.9 indicates that the majority of potential causes are associated with the re-winder itself and the type of product. In total, 21 'X's were identified, which formed the input of the data collection plan to construct the multi-vari study.

Data collection plan

Having established the current process performance and created a continuous project 'Y' metric, the Six Sigma team focused on understanding the relationships between changes in the downstream factors ('X's) and their impact on the outcome ('Y's). To achieve this, potential causes ('X's) of the project 'Y' were identified during the cause and effect diagram. However, for most of them there was no information available. A data collection plan was carried out to obtain this information. The data collection plan includes the 'X's to measure, their operational definitions, identification of data sources and data collection forms (Eckes, 2001).

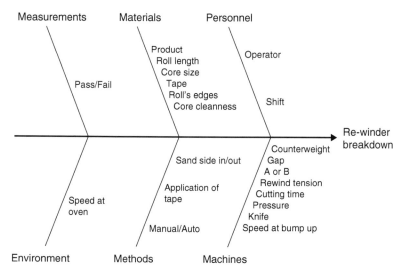

Figure 2.9 Cause and effect diagram for re-winder.

As an outcome of the measure phase, the Six Sigma team narrowed its focus on distinct groups of project issues and opportunities. After the completion of the measure phase the team achieved the following deliveries:

- Plan for collecting data that specifies the data type and collection technique.
- Validated measurement system that ensures R&R.
- Set of preliminary analysis results that provides project direction.
- Baseline measurement of current performance.

2.3.3 Analyze phase

The purpose of the analyze phase is to start learning about data in order to generate, segment, prioritize and verify the possible root causes and their relationship to the 'Y's or outputs (Waddick, 2001). The data collected from the measure phase served as an input for the analysis phase. As the data collection was carried out for a relative long period of time, it allowed the process to reveal its full range variation in the long-term basis.

Different tools and techniques were employed to find correlation between 'X's and project 'Y' in order to reduce the number of variables and select the 'vital few' for further analysis. Main effects plots were employed to log data means for different 'X's. The points in the plot in Figure 2.10 are the means of the response variable at various levels of each 'X's, with a reference line drawn at the grand mean of the response data (operation cycle time) (Antony, 2003). They were used for comparing magnitudes of the different 'X's on the response. It is important to mention that they are not the result of a design of experiment, but the result of the data collection in which different variables or factors were recorded on a wide range of levels.

Figure 2.10 shows a main effect plot for five different 'X's at different levels and their influence in operation cycle time. The result shown in Figure 2.10 confirmed that the knife or unit where the cut operation is performed seemed

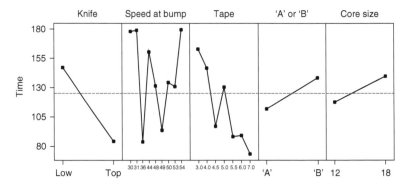

Figure 2.10 Main effect plot.

to be a significant contributor to the cutting time performance. Chop-overs on top unit (knife) average 70 units of time lower than the bottom unit. This corresponds to a better performance on the top unit (8.53% defect rate) in comparison to that of the bottom unit (10.55% defect rate). Investigation revealed that the difference in cut operational time between the bump and cut cycle on the bottom can be attributed to a possible difference in the pneumatic system in both units, the top unit having a more recent pneumatic controls. The team decided to update the pneumatics in the bottom unit to match those in the top unit. After the replacement of the pneumatic system a drop in the failure rate was achieved. However, it proved that there was more than one 'X' contributing to failures, since failures still occur.

The multi-vari chart shown in Figure 2.11 determines the interaction between the gap, spindle and core size variables. Multi-vari studies help associate key 'X's and the project 'Y,' identify noise variables and reduce the number of 'X's for the improvement phase. In addition, they are helpful in obtaining and understanding the process during its natural variation (Brue and Launsbry, 2003). The left-hand side of Figure 2.11 shows the process performance when the gap variable fluctuates for less than one standard deviation. As it can be seen, better performance (smaller the better) is recorded under this scenario. On the other hand, when this variable is outside one standard deviation the cutting operation time increases.

From the information obtained from the main effects plot and the multi-vari chart, the team decided to concentrate on the gap between new core (primed position) and the knife. It was thought that the gap between the new core and

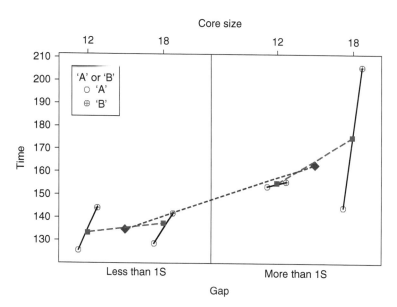

Figure 2.11 Multi-vari chart for time by 'A' or 'B' – gap core size.

the knife would be critical to the re-winder performance. If the gap was too small, the arm with the knife reaches the new core before the web is cut causing a failure. Conversely, where the gap is too large, the knife is able to cut through the web properly but the pack arm will fail to press the web onto the new core and so when the pack arm moves out of position the web moves from the core and a failure occurs due to lack of bond between new core and web. The theory is backed by the information shown in Figure 2.11. The maintenance team confirmed that the gap is determined by the stop position. A cam limit switch set up controls the re-winder turret position depending on the unit (top or bottom), spindle ('A' or 'B') and core size (12″ or 18″). However, inertia from the turret rotation and counterweight from the outgoing full roll may affect and modify this position after the limit switches system sends the stop signal. Further analysis was needed to determine that gap variable was now the critical 'X' and to establish the correct level setting of this subsystem CTQ characteristic.

Having carried out the data collection plan and analyzed the data with the aid of multi-vari charts, hypothesis test, gauge R&R, basic descriptive statistics and main effects plot, some 'X's were discarded for further analysis (e.g., operator, shift, pressure, line speed), and others were selected to form part of the vital few (gap, core size, spindle, unit/knife). As an outcome of the analysis phase, the Six Sigma team members had a strong understanding of the factors impacting their project, including:

- Key process input variables or the vital few 'X's that impact the 'Y.'
- Sources of variation (i.e., where the greatest degree of variation exists).

2.3.4 Improve phase

The improvement phase has the objective to consider the causes found in the analysis phase, and also select and target solutions to eliminate such causes. Multi-vari study helped to understand the relationship among different factors and conclude that the gap factor may be considered an important source of variation. Consequently, the team was primarily concerned about the capability and the specification limits of this subsystem CTQ characteristic.

Statistical tolerance

From the multi-vari study, the gap between the re-winder's pack arm and the new core has a significant impact on the outcome. Statistical tolerance of the gap was carried out to:

- determine the probability of re-winder failure based on the actual variation of the gap;
- identify the sources of variation in gap position and possible ways to eliminate them;

- understand what is an appropriate setting for the gap variable;
- estimate the process capability.

Subsystem CTQ sigma level

To establish the range of acceptability and capability of the gap, samples were taken including all possible combinations between core sizes (12″ and 18″), spindles ('A' and 'B') and re-winder's units (top and bottom). As a result, it was concluded that every time the gap is set above 1.200″ and below 0.350″, the machine fails. Having identified the upper and lower specification limits of the CTQ characteristic 'gap,' we can predict the process capability in the short and long term estimated by the sigma value. The sigma value indicates how often defects are likely to occur. The higher the sigma value, the less likely a process will produce defects (Harry, 1998). The short-term process capability is calculated from data taken over a short enough period of time that there are no external influences on the process (i.e., temperature changes, shift changes, operator changes, raw material, lot changes, etc.). Only, the so-called 'white noise' or 'common cause' variation produces changes in the response variable. However, degradation of the short-term performance of the process is largely due to the adverse effect of long-term external influences. This degradation result in shifts and drifts of the process which are generally not detected resulting in a different long-term capability. Contrary to short-term capability, the long-term capability is calculated from data taken over a period of time long enough that external factors can influence the process are present.

To calculate both capabilities, data was collected in 'rational subgroups' so that short-term variability can be assessed (limited by the technology in a process), coupled with long-term variability, and shifts and drifts of the mean (limited by technology and control of the process). Harry (2003) asserts that although the typical shift factor will frequently tend toward sigma 1.5 each CTQ characteristic will retain its own unique magnitude of dynamic expansion, which can be calculated using rational subgrouping. The 'Rational Subgrouping' technique quantifies variation by short-term variation (common cause variation) and shift/drift in the mean response (so-called 'assignable cause' variation, or 'black noise'). To this aim, data was collected in subgroups selected for the combinations of spindle ('A' or 'B') and core size (12″, 18″), and samples of 10 were taken consecutively to reduce the dispersion in the subgroup. In this manner, the team gives the maximum chance for the measurements of within subgroups to be alike and the maximum chance for between subgroups to differ one over the other. These subgroups were selected based on the cause and effect diagram, and the multi-vari study in which it is believed that different core sizes and spindles affect the gap. Then, the variation within subgroups or streams was calculated by the square root of the Mean Square Error Term of the analysis of variance (ANOVA) table. As a result, the sigma level short term was 3.52; whereas the sigma level long term was 1.47 sigma. Thus, the sigma shift which describes how well the process being measured is controlled over time is estimated to be 2.0499 (i.e., the difference between the sigma level short term and long term).

Gap's transfer function

Geometrically the gap in the re-winder is affected by three different parts: T1: the turret to reach the bump position; C2: core size and A3: bump unit or pack arm position (see Figure 2.12). From the component dimension and variations the geometric relationship between the components as the fit together to make an appropriate gap were estimated. This relationship is called transfer function. In other cases, the transfer function can be estimated by the linear regression using Design of Experiments. According to the transfer function, an increase in any of the components dimension decreases the gap. For example, if the core size increases the gap between the core and the knife or arm decreases. In addition, if the re-winder rotates clockwise (T1 increases) the gap decreases. Consequently, the system capability is made out of the capability of multiple parts and dimensions. Therefore, the parts capability dictates the system capability predicted previously; that is, variation in the parts results in variation on the outcome. Thus, the mean and standard deviation of each part was calculated as an input to the transfer function to predict the CTQ gap. As a result, it was concluded that between 74% and 88% of the variation comes from the turret (T1) factor. It was also noted that the turret's mean differs depending on core size (12″ or 18″) and spindle ('A' or 'B'). These results are consistent to those of the sample taken during the multi-vari study. On this, the combination of core size 18″ and 'B' turret produces the longest cutting operation time. This is followed by B-12″, A-18″ and finally A-12″. Therefore, the Six Sigma team concluded that the main source of variability is the turret part (T1). However, the current technology used in the re-winder was not capable to satisfy the specification limits. Nevertheless, through an understanding of the root causation of the problem the team was able to identify and implement an alternative solution during the control phase. The improved solution was implemented during the control phase, and described in the next section.

At the conclusion of the improve phase, the Six Sigma team achieved the following deliveries:

- Identification of alternative improvement.
- Validation of the improvement using transfer functions.

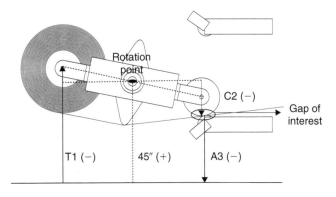

Figure 2.12 Vector diagram.

2.3.5 Control phase

Having identified the root causes of the problem and the possible solutions to reduce the variation of the process, the Six Sigma team moved to the control phase. This phase has the objective of implementing ongoing measures and actions to sustain the improvement by monitoring, standardizing, documenting and integrating the new process on a daily basis (Pande *et al.*, 2000).

The Six Sigma team identified a sustained solution capable of reducing variation in the turret position. However, the lead time for the implementation of this solution and the amount of waste produced daily led the team to implement a temporary solution. Both the solutions are described in the next section.

Temporary solution

The temporary solution involves the improvement of the gap by providing the operators with a feedback system, which allows them to understand the optimal turret position and to maintain that turret position. This solution consists of installing spirit levels on the extreme of the turret to measure the angle and therefore the gap between turret and pack arm. In this way the operator adjusts the gap according to the spirit bubble by moving the turret up or down until the bubble is centered between the lines printed on the spirit level. A camera was also installed for operators to observe the spirit levels from the machine controls. A total of eight levels were attached to the turret in order to satisfy all the combination of core sizes (12″ and 18″), units (top and bottom) and spindles ('A' and 'B'), which produce different gap.

Sustained solution

The sustained solution satisfies the overall objective by maximizing financial benefits, and improving process capability with a reasonable risk and investment. This solution involves a redesign of the turret indexing system. The current turret technology is not only incapable of achieving the quality levels required but are also at the end of their lifecycle. The new system is based on a new inverter to implement a controlled move to position, an absolute encoder to accurately measure this position and a disk brake to hold this location.

Process capability

Two years before the project started it averaged 1.29 sigma long term. The problem was faced using Six Sigma in order to reach the three failures per week goal, which is around 3% defect ratio (97% yield). As the project move forward, one of the root causes was identified and repaired by readjusting parameters of the machine. This produced an improvement of 5% to reach 6% defect ratio. After that the temporary solution was installed to reach around 2.7 sigma in the short term (2.06 sigma in the long term). This confirmed the strong relationship between the project 'Y' and the subsystem CTQ gap. Later the sustained solution was implemented improving even further to 2.64 sigma

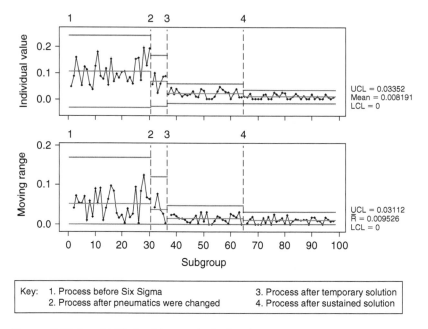

Figure 2.13 I and MR chart for re-winder by step.

long term. This can be perceived in the control chart in Figure 2.13 in terms of defect rates. In terms of costs of poor quality, the process waste costs were estimated at £60,720 per year. This does not include opportunity costs associated with running time. After the implementation of the temporary solution the costs were £11,246. However, the sustained solution reduced costs of poor quality at around £4276. This represents a reduction of £56,395 annually, or a yield improvement from 88.5% to 99.18%.

Control plan

Control plan is a set of documents that provides a point of reference among characteristics, specifications and instructions and links CTQs to the operational details of the process. It encompasses several process areas through operating procedures. The intent of an effective control plan strategy is to operate the temporal or sustained solutions consistently on target with minimum variation and minimizing process over-adjustment. It also helps to assure the identified and implemented process improvements become institutionalized by providing adequate training in all procedures.

The candidate variable to control is the gap, since the output variation is a function of the gap variation. A control plan, which indicates the target values, specifications limits and standard deviation expected for this CTQ characteristic, was put in place. The designated control method consists of an SOP with reaction plan. The Six Sigma team also developed a monitoring plan for the process in which the defect rate is plotted in a control chart weekly (see Figure 2.13). These charts, individual's chart (I) and moving range (MR) control chart, emphasize conformance to the specification of the process. In addition, it can be seen how the process has been improved after the different improvement steps were implemented.

Upon completion of the control phase, the process owner understood performance expectations, how to measure and monitor 'X's to ensure performance of the 'Y,' and what corrective actions should be executed if measurements drop below desired levels. The Six Sigma team made the transition of the process back to the process owner. This project was completed in the timescale established and was closed 8 months after the initialization date.

2.4 Conclusions

This chapter presented a real case study illustrating the effective use of Six Sigma to reduce waste in a continuous film line. It illustrates in detail how the project was selected, and how the DMAIC phases of the Six Sigma DMAIC methodology were carried out. Several tools and techniques were employed during the course of the project.

The success of this Six Sigma case study can be attributed to the following key factors:

1. Six Sigma methodology.
2. Management involvement and commitment.
3. Project selection and its link to business goals.
4. Training and teamwork.
5. Project progress tracking and monitoring.

The estimated savings generated from the project was well over US $120,000 US (approximately £60,000) per annum. In addition, the waste reduction created a chain reaction in which run time was increased, quality was improved and inspection reduced. Additional soft benefits were perceived, including employee participation in Six Sigma projects, increased process knowledge and the use of statistical thinking to solve problems. The rapid payback of the project motivated people at the company to implement more Six Sigma projects. Additional projects were selected based on the cluster of projects produced during the initial project selection exercise carried out during this project. Likewise, after completion of the project, the Six Sigma team members disband while the Black Belt begins the next Six Sigma Project with a new team.

Acknowledgments

This research has been funded by the Mexican Council of Science and Technology (CONACYT). Special thanks to Noel Byrne, Adrian Barnes and Karen Vyse.

References

Antony, J. (2003). *Design of Experiments for Engineers and Managers using Simple Graphical Tools*. Oxford: Butterworth/Heinemann.

Bañuelas, R. and Antony, J. (2002). Critical success factors for the successful implementation of Six Sigma projects in organisations. *The TQM Magazine*, 14(2), 92–99.

Berryman, M. (2002). DFSS and big payoffs. *Six Sigma forum magazine*, 2(1). On line at http://www.asq.org/pub/sixsigma/past/vol2_issue1/berryman.html. Retrieved on 12 December 2002.

Bicheno, J. (1994). *The Quality 50*. Buckingham: PICSIE Books.

Breyfogle, F. (1999). *Implementing Six Sigma; The Smarter Solutions Using Statistical Methods*. NJ: Wiley.

Breyfogle, F., Cupello, J. and Meadws, B. (2001). *Managing Six Sigma*. NY: Wiley Inter-science.

Brue, G. and Launsby, R. (2003). *Design for Six Sigma*. NY: McGraw-Hill.

Eckes, G. (2001). *The Six Sigma Revolution*. NY: Wiley.

Eckvall, D. and Juran, J. (1974). Manufacturing planning. In eds. Juran, J., Gryna, F. and Bingham, R. *Quality Control Handbook*. NY: McGraw Hill.

George, M. (2002). *Lean Six Sigma*. NY: McGraw-Hill.

Harry, M. (1994). *The Vision of Six Sigma*. Phoenix: Sigma Publishes Company.

Harry, M. (1998). Six Sigma: a breakthrough strategy for profitability. *Quality Progress*, 31(4), 60–64.

Huber, C. (2002). Straight talk on DFSS. *Six Sigma Forum Magazine*, 1(3). On line at http://www.asq.org/pub/sixsigma/past/vol1_issue4/dfss.html. Retrieved on 12 December 2002.

Ishikawa, K. (1974). *Guide to Quality Control*. Tokyo: Asian Productivity Press.

Kiemele, M., Schmidt, S. and Berdine, R. (1997). *Basic Statistics*. Colorado Springs: Air Academy Press.

Meredith, J. and Mantel, S. (2000). *Project Management*. NY: Wiley.

Montgomery, D. and Runger, G. (2003). *Applied Statistics and Probability for Engineers*. NY: Wiley.

Nave, D. (2002). How to compare Six Sigma, Lean and the theory of constraints. *Quality Progress*, 35(3), 73–79.

Pande, P., Neuman, R. and Cavanagh, R. (2000). The Six Sigma way: how GE, Motorola and other top companies are honing their performance. NY: McGraw-Hill.

Porter, L. (2001). Six Sigma excellence. *Quality Word*, 112–115.

Pyzdek, T. (2001). *The Six Sigma Handbook*. NY: McGraw-Hill.

Rasis, D., Gitlow, H. and Popovich, E. (2002). Paper organisers international: a fictitious Six Sigma green belt case study. I. *Quality Engineering*, 15(1), 127–145.

Shephered, F. (1994). *Modern Coating Technology Systems*. Barnet, UK: Maclaren.

Waddick, P. (2001). Six Sigma DMAIC quick reference. On line at http://www.isixsigma.com. Retrieved on 16 August 2003.

3

Process improvement at Tata Steel using the ASPIRE DMAIC approach

D.P. Deshpande, Supratim Halder, Suman Biswas, Arun Raychaudhuri, Asim Choudhary and Ashok Kumar

3.1 Tata Steel Company (TBEM 2004)

With a turnover of approximately $2.5 billion, Tata Steel is Asia's first and India's largest integrated private sector steel company. Over the years, it has consistently overcome the challenges of a highly competitive global economy and has, 'committed to become a supplier of choice by delighting its customers with service and products.' For the past 10 years, Tata Steel follows its own Tata business excellence model (TBEM) to seek and identify opportunities for improvement, and bring about excellence in all its endeavors, while constantly benchmarking (BM) its performance with the best in the world and always striving to achieve or exceed these benchmarks. Among the core values of Tata Steel are improvement orientation, participative management, innovation, role model leadership, and focus on customers. At the current time, Tata Steel is an equal opportunity employer with a payroll of nearly 41,211 and through diligent planning, has created a cosmopolitan mix among its employees. Using TBEM as a change driver, Tata Steel has put together a strategic management system in place that constantly measures strategic variables (productivity, quality, innovation, agility, societal welfare, community service, and others) through balanced scorecards and other instruments. In terms of customer focus, business results, and other strategic processes, Tata Steel measures and monitors some 150 different performance measures. In order to ensure that the organizational output is consistent with Tata's vision and mission, Tata Steel has developed an exhaustive structure of process control and improvement within the TBEM framework. Twelve key processes, such as leadership, strategic planning and risk management, market development, investment management, improvement and change management, etc., are linked with key value creating processes and key support processes which are monitored closely through appropriate yardsticks set by business excellence model. The results are impressive enough to keep Tata Steel among world leaders and reflected in Tata Steel's accomplishments and awards. Among many distinctions, Tata Steel won Asia's Most Admired Knowledge Enterprise award and National Association of Software and Service Companies (NASSCOM) award for best information technology (IT) user in 2003.

3.2 ASPIRE and Six Sigma at Tata Steel

Until 2002, Tata Steel was following a modular framework that used numerous tools and techniques to bring about process improvements in a contextual fashion but lacked a systemic approach. Numerous process/performance improvement techniques and models, such as VE, ISO, QIP, TBEM, etc., were adopted over the years (see chart in next page). This type of application did not have a perspective or organized framework that would guarantee that the problem solutions developed were the best or most optimal ones. Also, due to the localized and contextual nature, the leverage of solutions developed for one problem was not available for other similar problems elsewhere in the organization. In order to address this issue and provide an integrated framework, Aspirational Initiatives to Retain Excellence (ASPIRE) program was conceived in June 2002. ASPIRE program was started with Six Sigma methodology, taking some of the pilot projects. After successful deployment of Six Sigma, it was merged with other initiatives like total productive maintenance (TPM), 5S, value engineering (VE), Poka-Yoka, theory of constraints and named as ASPIRE.

As a change driver, TBEM is perpetually seeking to improve organizational performance in all its processes and functions. A key differentiating factor (compared to the practices in other industries) of this approach is that the search for improvements in processes is not limited by resource availability. Indeed, the search for excellence is driven by the restlessness and frustration with the *status quo* and once new performance improvement opportunities are identified, resource availability becomes a management prerogative. The incessant drive for breaking away from the *status quo* and developing better processes feeds into a 'virtuous cycle' of ambition. This virtuous cycle is formalized as the underlying philosophy of ASPIRE. Under this philosophy, the trigger for improvements need not be external (e.g., cost of poor quality or opportunities for improvement). In June 2002, Tata Steel conceptualized, created, and communicated ASPIRE as an integrating framework of improvement techniques that could help accelerate the pace of change. Numerous modular programs, including Six Sigma, that were earlier used for obtaining process improvements through program-specific approaches, were clubbed under the umbrella of ASPIRE, as stated above. See figure in next page.

On 24 June 2002, ASPIRE made its maiden attempt to create value for customers and channel partners through customer value management (CVM), retail value management (RVM), and customer product optimization (CPO). Subsequently, the same concepts were applied for creating value with suppliers through a supplier value management (SVM) program.

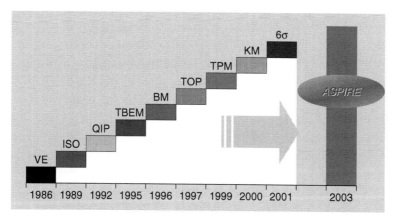

Legend: VE = Value Engineering
         ISO = International Organization for Standardization
         QIP = Quality Improvement Projects
         BM = Benchmarking

TOP = Total Operational Performance
TPM = Total Productive Maintenance
KM = Knowledge Management
6σ = Six Sigma

We now illustrate how the ASPIRE DMAIC methodology (DMAIC, Define–Measure–Analyze–Implement–Control) was deployed at Tata Steel to obtain significant process improvements. The coke moisture project was one of the pilot projects of Six Sigma initiative of the company. The success of the project was very critical for Six Sigma deployment in Tata Steel as well as for the improvement in the process of coke moisture control.

3.3 Project selection

While the project selection process precedes a project's define phase, there is a chicken and egg relationship between identification of an initial general goal and the 'define phase,' as that goal is better understood and refined over first few deliberations. Usually one has an initial idea of one's goal, but without execution of some of the define work, one does not know if the scope is reasonable. Project selection brings out another important consideration not directly addressed by 'define' step. It establishes the link between candidate projects and corporate strategy (in one sense, these are the top level Ys – see Figure 3.1).

3.3.1 Introduction to the project

In a typical steel plant like Tata Steel, blast furnaces play a major role. In blast furnaces, iron ore is reduced to hot metal (liquid iron) with the help of coke. Coke provides the sensible heat and heat of reduction in the blast furnace. The hot metal is then converted into crude steel and subsequently rolled into hot rolled coils, wire rods, etc.

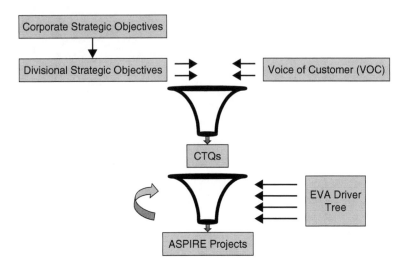

Figure 3.1 Project selection.

Typically for the production of 1 ton of hot metal 400–500 kg of coke is required. Coke is also the most expensive raw material in the blast furnace operation. In view of the volume of this material and its cost, it is natural to expect that coke is used in at cost-optimal fashion and therefore it must comply with stringent quality specifications. One of the important coke quality parameters that influence the coke demand in the blast furnace is the coke moisture and its consistency. Fluctuations of the moisture level are reflected as fluctuations in the energy supply to the blast furnace thereby disturbing the blast furnace process and also wasteful use of this expensive energy source.

Coke is produced in the coke ovens by carbonization of metallurgical coals. The carbonization process operates at a temperature of about 1000–1100°C. The product of the process is red-hot coke coming out of the ovens at such a high temperature. The coke so produced is pushed out of the ovens and collected in cars known as quenching cars.

3.3.2 Purpose of coke quenching

The purpose of coke quenching tower is to quench the red-hot glowing coke in the quenching cars by intense water sprays, preventing it from getting burnt out. This operation unfortunately leaves some residual water in the quenched coke. The quenching process is designed to produce coke with the least amount of water. This means that the quenched coke will still be warm enough to evaporate the residual water left in the quenching process, leaving the coke as dry as possible. The quenching also has been so effective that the coke does not burn (the combustion temperature of coke is about 400 °C). On the other hand, if the coke is not adequately quenched, then it needs further local quenching on wharf to prevent damage to the downstream conveyer belts.

3.3.3 Quenching process

Typically, a quenching tower is a reinforced concrete frame with brick line structure. The quenching car carrying red-hot coke moves into this quenching tower. The water used for quenching is recycled after settling in the breeze pond and pumped up to the overhead tanks of the quenching tower. The quenching towers provided at the quenching station of coke plant have technologies of different states of development. Battery 3, the only top charged battery in Tata Steel, has a quenching tower of design of 1970s. It has four large-size spray nozzles, laid to cover the quenching car. At Batteries 5, 6, and 7, there are many quenching nozzles, each giving a laminar downward flow, yet covering the entire quenching car. At Batteries 8 and 9, a semi-flood quench system is employed, which means the water sprayed on the quenching car is not allowed to drain away creating a pool of water in the quenching car. The coke floats in the car, levels itself, and gets quenched more uniformly and instantly.

Semi-flooded quenching means large quantity of water sprayed onto the quenching car initially for a short period of time to create flooding up to certain height. The generated steam quenches the upper layers of coke in the quenching car partially to be followed by water quenching. Quick flooding requires faster flow of water. Therefore, few additional nozzles are provided on the side of the quenching car at an angle of 45° with horizontal. The side quenching operates for a few initial seconds during the process. After the lapse of a predefined time, the side quenching stops and top quenching starts, which creates a laminar flow of water to permit full quenching of coke for a specified time. The total quenching time is kept approximately 90 sec as per the basic design, but can be adjusted as per the requirement. Thus this process achieves a uniform and full quenching of hot coke.

The quenching water comes from fixed head overhead tanks through motor operated quick acting butterfly valves. Each circuit for top quenching and side quenching is provided with two valves, so that one valve of each circuit can be kept in a standby position.

A need was felt to bring about improvement in the average levels of residual coke moisture achieved as well as keeping its standard deviation low. A large number of improvement projects taken in the past had helped in improving spraying water quality, the reliability of equipments, the response time of valves, etc. But these projects had not been able to make a permanent impact on the intended objectives. Therefore, this process was chosen for ASPIRE DMAIC approach for improvement.

3.4 ASPIRE DMAIC approach

3.4.1 Define phase

The first phase of a DMAIC process improvement project, known as 'define', has three steps:

1. Identify the customer(s) and the 'critical to quality' (CTQ) requirement that will be the focus of the project. The CTQ may also be referred to as the project 'Y' [as in $Y = f(x)$].

2. Create the project charter.
3. Develop a high-level process map.

As a general rule, the scope of an ASPIRE DMAIC project is limited so that the project can be completed within 4–6 months.

Identification of CTQ

The project started with a very strong voice of customer (VOC), which in turn was directly linked with the corporate strategic objective. One critical strategic objective of the company was 'cost reduction' and cost of hot metal produced from blast furnaces was one of the most important components of the cost of steel. Cost of hot metal is directly linked with the amount of coke used in blast furnaces and this depends on the moisture content in coke. Therefore, the blast furnaces always prefer to have coke with lower coke moisture and they insist that moisture content in coke should never be more than 3.5%. The project started with this VOC and naturally the CTQ identified was 'moisture % in coke.'

Project charter

The typical components of project charter were then worked out as follows:

(a) *Business case*: One of the important coke quality parameters is the coke moisture and its consistency. Fluctuations of the moisture level are reflected as fluctuations in the energy supply to the blast furnace thereby disturbing the isotherms in the process. The higher moisture level or its fluctuation leads to an increased demand for coke. Since the quenching process introduces moisture in coke; it is only logical to examine this process to bring about improvements.

(b) *Problem statement*: Most of the steel plants around the world have adopted wet quenching method and they maintain the moisture level below 4% with tight controls on its variation. At Tata Steel coke moisture was targeted at 3.5%. In spite of making several attempts to reduce coke moisture over the past 10 years at Tata Steel, it was not possible to control it consistently below 3.5%.

(c) *Goal statement*: The goal was taken as to get the coke moisture value consistently below 3.5%, with more than 50% reduction in the standard deviation.

(d) *Project team*: The project sponsor (champion) is the Vice President (VP) of iron making, and team members include Mr. D.P. Deshpande, Mr. Supratim Halder, Mr. Suman Biswas, Mr. Arun Roychaudhuri, and Mr. Asim Choudhary.

(e) *Scope*: The project will concern with coke moisture and its standard deviation as seen at the outlet of quenching tower only, not concern itself with the effect of other extraneous water addition as received in the rainy season.

(f) *Financial opportunity*: It was also theoretically calculated that a 1% improvement in the average value of the coke moisture leads to a monetary benefit of US$4.1 millions per year. Thus improvement of the coke moisture values has a significant impact on the bottom line.

High-level process map

Supplier	Input	Process	Output	Customer
Coke oven batteries	Hot coke having temperature of 1000°C	Coke quenching process	Wet coke	Blast furnaces

3.4.2 Measure phase

The measure phase can be viewed as consisting of three steps:

1. *Confirm/refine the project Y (CTQ)*: Coke moisture improvement was a big challenge for last few years as shown in Figure 3.2.
2. *Define performance goals or standards*: Measurement of coke moisture is carried out every shift in a laboratory by recording the loss in weight of the sample during testing. For this purpose, a known weight of sample is collected through an auto sampler located at the discharge of the coke conveyor of the coke handling plant.

Project Ys	What to measure	UOM	Target	LSL/USL	Defect definition
Coke moisture	Moisture % present in coke	%	3.0	USL = 3.5	Any data point above 3.5

UOM: unit of measurement; LSL: lower specification limit; USL: upper specification limit.

3. *Calibrate the measurement system*: Since this measurement system is not automated, the team performed a Gage R&R study to validate the measurement system and found out that the repeatability and reproducibility of the system are within the limits.

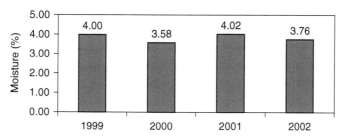

Figure 3.2 Average coke moisture from year 1999 February to 2002 February.

3.4.3 Analyze phase

1. *Measure the capability of the existing process*: The USL of coke moisture was defined as 3.5% considering the 'VOC' from blast furnaces. As the distribution of the coke moisture data is not normal, it was decided to calculate the sigma level following the DPMO method.
2. *Calculate the sigma level*: In the case of coke moisture, the defect percentage has been 44.5 (i.e., DPMO = 445,000), which corresponds to a baseline sigma level of 1.6. This is not an acceptable performance, and a target was taken to improve the sigma level of moisture up to 3 (corresponding to 6.7% defect). The detailed process map of coke quenching process is shown in Figure 3.3.
3. *Identify possible Xs*: To improve the coke moisture at Batteries 8 and 9, some problem areas were identified after brainstorming and these were analyzed with the help of a 'failure modes and effects analysis' (FMEA).

   The FMEA is shown in Table 3.1.

   The FMEA clearly triggered some action points. Some of them were obvious and easily implementable, and they were taken care of immediately by the team. Each of these problems was solved by a scheduled time and giving responsibility to concerned person. The problem areas with their solutions have been listed below:
   - Difference in pressure head makes the flow rate inconsistent; therefore, the water level in the quenching tank before quenching was maintained at a maximum and constant level (99.9%) on a regular basis.
   - No leakage of water was ensured from dry fog nozzles, quenching car flap sealing, and from water spraying valve in the conveyor circuit.
   - The incoming source of coke piece from quenching station was stopped to avoid mixing of the swamp breeze.
   - Double quenching in the quenching station was reduced.

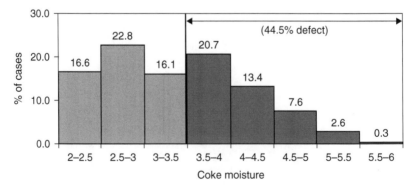

Figure 3.3 Distribution of coke moisture values before experimentation.

Table 3.1 FMEA performed on quenching

Process/operation/ function	Failure mode	Effect of failure	Severity	Causes	Occurrence	Control	Detectability	RPN
Quenching	Inconsistency in flow rate	Different amount of water used	4	Different water level in quenching tank	7	No control	8	224
	Leakage of water	Extra water added	6	Leakage from dry fog nozzles, flap sealing, spraying valve	5	Time-to-time maintenance	5	150
	Double quenching/ spot quenching	Excess moisture in coke	8	Initial quenching not proper, faulty parameter design	9	By experience	6	432
	Non-uniform distribution of coke on quenching car	Hot coke in the wharf	7	No synchronization of speed between loco and pusher ram	8	Controlled by skill level of operator	7	392
	Water carry over in the wharf by quenching car	Higher moisture	7	Improper drain-out, improper parameter design	5	By experience	6	210

- All the valves were kept in the fully open condition to maintain the required speed of the water in the quenching circuit.
- Spot quenching of the hot coke on the wharf was minimized.

Marginal improvement in the coke moisture could be seen with the implementation of the steps listed above. However, the other variables that came out of the FMEA exercise required further analysis and they were carried forward to the next phase.

4. *Identify and verify critical Xs*: The problem was discussed among the experts on the subject and all operational and other factors that impact the moisture level in coke were identified. After a detailed analysis, three critical parameters were shortlisted. The parameters are:
   - Quenching time (amount of time used to quench the hot coke in the quenching car).
   - Crack opening time (time up to which flooding of water in the quenching car is allowed).
   - Drain-out time (amount of time used to drain the water out from quenching car).

As there was no established relationship between the above listed process parameters and coke moisture, the control by the operators was based on gut feeling and hunches. Since, enough past data was not available for analysis with reliability, real time experiments were planned using design of experiment (DOE).

3.4.4 Improve phase

Improve is a three-step process which we will describe below.

1. Identify solution alternatives

Designing a systematic experiment   Once the parameters (for experimentation) were finalized, the challenge was to find out what should be the optimal combinations of these parameters for getting lower average value and standard deviation of the coke moisture. The previous values of these parameters were (a) quenching time – 85 sec, (b) crack opening time – 55 sec, and (c) drain-out time – 180 sec. It was decided to take three levels for the selected parameters. For quenching time, three levels were 75, 85, and 95 sec; for drain-out time, they were 160, 180, and 200 sec; and for crack opening time, they were 55, 65, and 75 sec. This was a three-level three-factor experiment (full factorial design). Total number of runs were 3^3 (i.e., 27). The change in coke moisture with the change of these operating parameters is believed to follow a complex pattern, and a non-linear trend was expected to come up. Therefore, the team decided to go for a three-level DOE. Also, each single 'run' of this experiment was supposed to take a couple of hours and the cost of such experiments (if moisture value goes high) would also adversely impact the customer (blast furnaces). Therefore, it was decided to take only one response for each combination (no replication). But for each single run, three values at three different pre-determined time-points were taken (repetition). The combinations for levels and factors have been shown as follows:

Factor	Level		
	1	2	3
Quenching time (sec)	75	85	95
Drain-out time (sec)	160	180	200
Crack opening time (sec)	45	55	65

Another alternative solution (an innovation)   Since all of the above parameters are related to the amount of water being used in quenching of coke, and also since it is obvious that the water used for quenching is a predominant source of moisture content in coke, an innovative approach was conceived to reduce the amount of water used for quenching.

During the experimentation, we felt that the temperature of re-circulating water in the quenching station could be an important parameter for reducing coke moisture. The hot coke is cooled by water in the quenching station. In the existing system, the water is re-circulated to quench the hot coke. The temperature of re-circulating water is around 60°C. The heat balance equation, assuming heat given out by the hot coke is taken up by the water, leads to:

$$M_c S_c (T_{IC} - T_{FC}) = M_W S_W (T_{FW} - T_{IW}) + E \times M_W \times L_W$$

where
M_c mass of hot coke
S_c specific heat of coke
T_{IC} initial temperature of hot coke (1000°C)
T_{FC} final temperature of hot coke after quenching (200°C)
M_W mass of water used to quench the hot coke
S_W specific heat of water
T_{FW} final temperature of water used for quenching (60°C)
T_{IW} initial temperature of water after quenching (100°C)
E % of evaporation loss during quenching
L_W latent heat of evaporation for water.

Assumptions
1. Heat taken by the superheated steam is negligible and has been ignored in the equation.
2. Heat transfer by the quenching loco, which carries the hot coke to quenching station, has been ignored.
3. Mode of heat transfer in the quenching station has not been considered.

Now, from the heat balance equation, making the left-hand side of the equation as constant, if we change the initial temperature of the water from 60°C to ambient temperature (30°C), the mass of water used for quenching can be reduced by 20%. If the flow rate is constant, the quenching time can be reduced to quench same amount of hot coke. As quenching time plays a major role in the control of the coke moisture, by reducing quenching time, the coke moisture can be reduced to a great extent.

Validation of the idea Following the idea, experiments were designed and carried out. To get ambient temperature of water, service water was used for experimental purpose. The coke moisture value was recorded below 2.0%. For the first experiment, the average value of coke moisture was recorded as 1.7% and for the second experiment, it was 1.95%. Most importantly, there was no hot coke on the wharf during the experiment.

2. Select the best solution and the relationships between Xs and Ys

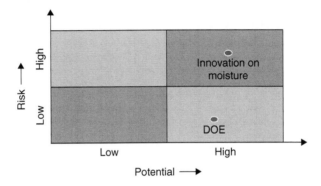

Though the innovative idea had a very high potential, the investment cost to implement the idea was very high and the validation was not done for longer period of time. The project team decided not to implement the idea immediately and park it for future project. It was obvious to plan a DOE to get the best possible solution.

After finalizing the design layout, every combination was experimented for 1 h and the readings of coke moisture were taken in Nucleonic Moisture Gauge (NMG) in coke control. The experimentations were carried on for 4 days. The interactions and main effects of the parameters were plotted and two equations on average and standard deviation of the coke moisture were arrived at.

$$\text{Average coke moisture} = 2.87 - 0.22 \times X + 0.04 \times Y - 0.04 \times Z + 0.01 \times X \times Y - 0.45 \times X \times Z - 0.03 \times Y \times Z - 0.12 \times X^2 + 0.1 \times Y^2 + 0.09 \times Z^2$$

where
X (quenching time $-$ 85)/10
Y (drain-out time $-$ 180)/20
Z (crack opening time $-$ 55)/10.

With the equation, the average coke moisture value was predicted and R^2 value was computed to be 75%. Similar equation on the standard deviation of coke moisture was developed as a function of the quenching time, the drain-out time, and the crack opening time.

3. Implementation plan

With the help of the above equation, three possible two-way contour plots were developed to find out the optimal combination of the parameters, which would result in the lowest coke moisture (see Figure 3.4). From the contour plot of quenching time vs. drain-out time, it was seen that for quenching time of 75 sec and at a range of 160–190 sec drain-out time, it was consequent to the lowest level of the coke moisture. Again it was seen from the contour plot of crack opening

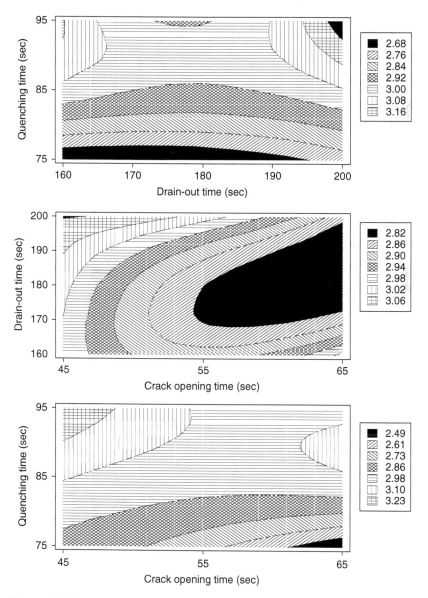

Figure 3.4 Two-way contour plots of coke moisture (batteries 8 and 9).

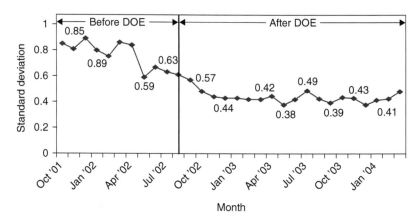

Figure 3.5 Standard deviation in coke moisture before and after DOE.

time vs. drain-out time, that the lowest coke moisture could be obtained for a range of crack opening time of 55–65 sec and a drain-out time of 170–200 sec. The quenching time and drain-out time were fixed as 75 and 180 sec, respectively. The contour plot for crack opening time vs. quenching time showed that the least coke moisture could be obtained for a quenching time of 75 sec and a crack opening time of 65 sec. Combining all the above observations, it was concluded that the lowest coke moisture could be achieved for the following values of the parameters:

- quenching time: 75 sec
- drain-out time: 180 sec
- crack opening time: 65 sec.

3.4.5 Control phase

The purpose of the control phase is to make sure that our improvements are sustained and reinforced. We want to be sure we put in place all of the actions that will help the change be both successful and lasting.

Control can be described as a three-step process:

1. *Develop control plan*: After implementing the best possible combination of vital Xs, the challenge was to make the performance of coke moisture consistent. Control charts (I and MR chart) used to monitor coke moisture results.
2. *Determine improved process capability*: After implementing the optimal values for the parameters, the average value of coke moisture improved from a base value of 3.43–2.7%. Standard deviation of coke moisture also improved from 0.9 to 0.44 (see Figure 3.5). Sigma level of coke moisture was monitored based on shift wise coke moisture values (considering any value >3.5% as a defect), which also improved from a base level of 1.6–3.3 (see Figure 3.6).
3. *Implement process control*: To sustain the improvement in the long run, rigorous control mechanism was followed to maintain the parameters at the

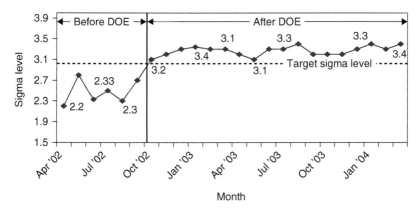

Figure 3.6 Sigma level of coke moisture before and after DOE.

optimal levels (as obtained from the experiment). These levels are maintained in the two-level automation system and no manual intervention (to change the parameter levels) is allowed any more.

3.5 Similar attempt at Batteries 5, 6, and 7

Encouraged by the success at Batteries 8 and 9, the DOE was extended to Batteries 5, 6, and 7. This required a two-factor three-level full factorial DOE.

The factors were quenching time and drain-out time. The design matrix has been shown in the following table:

Factor	Level		
	1	2	3
Quenching time (sec)	90	100	110
Drain-out time (sec)	120	130	140

Total number of runs were 3^2 (i.e., 9). After analyzing the results, the best results were seen for a quenching time of 95 sec and drain-out time of 140 sec. The optimal values of the parameters were implemented and there was significant improvement in the coke moisture. Average value improved from 3.4% to 2.78% and the standard deviation reduced from 0.96 to 0.31. Sigma level of coke moisture also improved from the base level of 1.6–3.1.

3.5.1 Solving the hot coke problem

The temperature of the hot coke in the wharf was an undesirable 200°C. If the temperature of the coke is high, it may damage the conveyor belts, which means loss of production. A thorough investigation of the problem revealed that the distribution of the hot coke in the quenching car was not uniform. During the pushing operation the synchronization speed of the quenching loco with the pusher ram was controlled by a tachometer, which was installed with the long travel drive motor shaft of the quenching loco. Since the weight of the quenching loco is 45 ton, whereas the weight of the quenching car with coke is 145 ton, the quenching loco slippage during pushing was not unexpected. Due to slippage a wrong feedback was sent to the pusher ram for the physical movement of the quenching car.

A brainstorming session was conducted to solve the problem. The feedback system of the quenching car movement was modified and installed with the quenching car travel wheel. Thus the feedback of quenching car travel with the physical movement is now in complete synchronization thereby solving the problem of hot coke distribution onto the quenching car.

3.6 Conclusion

This chapter presents an application of DMAIC methodology within Tata Steel based in India. One of the important coke quality parameters that influence the coke demand in the blast furnace is the coke moisture and its consistency. For the last 10 years, several attempts were made to reduce the coke moisture of Batteries 5, 6, and 7, and as a result, the coke moisture value came down to around 3.5%. The value remained stabilized at that level for the last 3 years and it was not possible to bring it down further with the conventional quality tools. A proper application of DOE resulted in a breakthrough improvement. This methodology can be replicated in other areas also, thereby creating possibilities for additional improvement. Batteries 8 and 9 started in 1998, and the average coke moisture value here was around 3.5% since its inception. After the experimentation the value improved significantly. A critical process parameter like coke moisture, which has a high bottom line impact, shows world-class results when it is optimized using DOE. The distribution of coke moisture values was improved significantly. Only 3.3% of coke moisture values were falling above 3.5%, as per the requirements of blast furnaces, whereas in previous case it was 44.5%.

Acknowledgment

We are thankful to Mr. Bijan Sharma, Mr. P.M. Kundu, Mr. S.C.P. Singh, Mr. N. Dutta, Mr. N.K. Bhalla, and all the foremen and operators of the quenching station for assisting us with the experiments.

References

Garwin, D.A. (1992). *Operations Strategy: Text and Cases*. New Jersey: Prentice Hall.
Halpin, J.F. (1966). *Zero Defects: A New Dimension in Quality Assurance*, ASIN: B0006BO4HS. New York: McGraw-Hill.

Hunter, W.G. (1975). *101 Ways to Design an Experiment*. Technical Report Number 413, University of Wisconsin, Madison.

Montgomery, D. (1991). *Design and Analysis of Experiments* (3rd Edition). John Wiley & Sons.

Pande, P.S., Neuman, R.P. and Cavanagh, R.R. *The Six Sigma Way – How GE, Motorola, and Other Top Companies are Honing Their Performance*. New York: McGraw-Hill.

Swamidass, P.M. (1986). Manufacturing strategy: its assessment and practice. *Journal of Operations Management*, 6(4), 471–484.

4

Improving product reliability using Six Sigma

Nick Shubotham, Ricardo Bañuelas and Jiju Antony

4.1 Introduction

This case study was carried out at one of the largest European white goods producers. The company where the case study was performed has successfully implemented Six Sigma since 1998 and has focused its efforts on reducing reliability problems. These problems are usually manifested in the field when the product fails to perform according to the design intend under the specific conditions for the desired period of time, causing customer dissatisfaction, increasing service and warranty costs and decreasing sales. In this chapter, the authors illustrate the use of Six Sigma DMAIC (Define–Measure–Analyze–Improve–Control) methodology in a step-by-step fashion in order to improve the reliability of tumble dryers. The chapter starts by introducing the Six Sigma practice in the company and describing the characteristics of the white goods industry. This is followed by the selection of a project aimed at improving reliability of tumble dryers. The chapter then illustrates the different phases of the DMAIC methodology and the tools and techniques employed to narrow down the project and identify the sources of variation. With the use of DMAIC methodology a new technology was selected, tested and introduced to reduce the main cause of reliability problems. As a result the savings generated from this project were estimated at around £335,000 annually. The authors conclude the chapter by offering a brief summary and stating the lessons learned during the course of this project.

4.2 Six Sigma in the collaborative company

By the time Six Sigma was introduced in the company, customers were demanding more and competitors were also improving quality. In addition, the costs of poor quality were estimated at around 15–20% of revenue. Based on the successful implementation of Six Sigma in other companies, the company decided to launch Six Sigma as a catalyst for focusing on customer satisfaction. Since then, Six Sigma has been seen as an efficient way to solve problems and make decisions as well as an opportunity to accelerate the company's performance and leverage best practices from the industry. To achieve a successful implementation of Six Sigma employees are organized in cross-functional teams. Team members are trained in statistical analysis, problem solving and leadership. Six Sigma leaders reside in every department and every site of the organization. Team members have different roles and responsibilities in the team depending upon their position in the company, forming a team of Champions, Master Black Belts, Black Belts, Green Belts and process owners. They focus their efforts on the customers' and/or consumers' satisfaction. Customers could be external or internal. External customers are businesses/organizations that receive appliances but do not use the appliances themselves (e.g., retailers, wholesalers and distributors). Whereas internal customers are functions that add value to a product or process for the benefit of an external customer/consumer (e.g., distribution is a customer of the warehouse). On the other hand, consumers are the end users of the product and are usually considered external (e.g., homeowner). To improve customer/consumer satisfaction the DMAIC Six Sigma methodology is employed throughout the organization. It provides with a data-driven methodology for achieving sustained process improvements by reducing defects. This methodology is used to identify inadequacies in the existing processes and to make lasting and controllable changes to products and processes that improve product quality, customer satisfaction and the company's profitability. DMAIC can be applied to manufacturing, design or commercial quality. As a result of Six Sigma implementation, the company has produced significant financial results as shown in Figure 4.1. In addition the following benefits as a result of the Six Sigma implementation have been perceived:

- consistent measurements and analysis methods across the company;
- better understanding of the process capability and the critical to quality characteristics (CTQs) that affect consumer/customer satisfaction;
- improved quality of products/processes;
- fewer defects, lower service call rates (SCR);

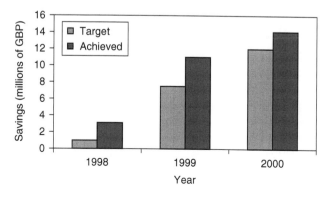

Figure 4.1 Six Sigma financial benefits.

- reduced operational costs;
- understanding of factors influencing company growth.

During this chapter an example of a Six Sigma DMAIC project carried out in the company is presented. This project was targeted to the consumer by improving the reliability of a family of tumble dryers produced by the company. In addition, by improving reliability, it was also possible to reduce the number of parts in the product, reduce labor and reduce the amount of money tied to inventory. In the company all Six Sigma projects must be targeted for process and product improvements that have direct impact on both financial and operation goals. Therefore, before presenting the DMAIC project in a step-by-step fashion, the business case for the project is introduced.

4.2.1 The business case

The business case intends to state the necessity of carrying out the Six Sigma project in order to satisfy customer needs and support business strategy. It helps to create management involvement, participation and authorization of the project. To this end, it is important to understand the industry, the company and above all the customer's needs. The voice of the consumer is usually gathered using market research which aims at discovering key factors when purchasing a tumble dryer. According to such researches, reliability is a key factor influencing purchasing decisions. Generally it is those consumers in the older age groups who are more likely to be concerned about reliability. Families are generally the heaviest users of their laundry appliances and so reliability is also of great importance. The company aims at providing reliable goods to customers and continually monitors product reliability using as a metric the SCR. An examination of the SCR issues reveal that no drum drive caused by belts is one of the top issues. Belts in tumble dryers are necessary to transmit movement from the motor to the drum and achieve an even and a fast drying. The current method of ensuring sufficient belt tension is by the addition of jockey wheel assemblies and springs. With this method, as the belts wear the tension is kept constant by the movement of the jockey pulleys. However, jockey pulleys can seize causing excessive loads on the belt resulting in no drum drive. Over the last 12 months failed belts have cost of £19,044. In addition, each failure causes customer dissatisfaction which can produce a decrease in sales. An alternative technology is to use elastic belts and avoid these problems. However, the use of elastic drive belts on dryers has been attempted several times by different engineers and different suppliers and all have ended in failure. The reasons for failure have been varied, including insufficient tension to drive the drum; excessive tension, making it difficult to fit; and inconsistent tension due to excessive part variation.

Having failed in introducing elastic belts in the past, the challenge is to overcome the mind set that elastic belts do not work by identifying the key stakeholders' issues and systematically proving the viability and robustness of the new drive system using Six Sigma. The proposed method is to use belts of an elastic construction. These are installed with a higher initial tension but relax quickly to a steady tension and remain close to this level throughout the life of the belt. The potential advantages of this method are:

1. Product cost reduction of approximately £240,000 per annum.
2. Reduction in the number of parts used (7 parts in total).
3. Reduction in material cost of the product due to the reduction in the number of parts.
4. Reduction in labor.
5. Reduction in SCR.
6. Less parts to develop faults.
7. Improved life of the drive belts.
8. Quieter operation because of the deletion of the jockey assemblies.

Given that this method is financially justifiable and has a direct impact on the consumers, the company launched a Six Sigma project which aimed at identifying, quantifying and eliminating the source of variation that leads to belt failure in tumble dryers. The top management gave the adequate support needed for the completion of, and permanent change to, the process by assigning a group of Black Belts and specialists to carry out the project. They also periodically reviewed and participated in the progress of the project.

4.3 DMAIC case study

4.3.1 Define phase

The define phase aims at identifying the customer and their CTQs, developing and refining a team charter, and mapping the business process to be improved. The following sub-steps were performed during the define phase of the DMAIC methodology:

1. define the problem,
2. define the advocacy team,
3. quantify potential financial benefits,
4. create a project charter.

Define the problem

A major cause of futile attempts to solve a Six Sigma project is a poor up-front statement of the problem. To overcome this issue, an initial 'as is' problem statement condition was clearly stated. This statement describes the problem 'as is' at the beginning of the project. It is a statement of what needs to be improved or what concerns the shareholders. The statement should contain data-based measures of the issue. In this case, the 'as is' situation is:

> *Current SCR for 5 kg tumble dryers is 7.25% annually. An important issue of high SCR is no drum drive caused by poor tensioning in belts. Tumble dryers currently use jockey pulleys as a means of tensioning the drive belt. Over the last 12 months failed belts have cost of £19,044.*

Having stated the 'as is' situation, the desired state was described. 'Desired state' is a description of what the project wants to achieve by improving the 'as is' situation. As the 'as is' statement, the 'desired state' is based on data. The desired state in this case is:

> *To improve reliability and reduce the product cost by removing the need for secondary belt tensioning. This desired state gives a potential saving of £316,000. Additionally, the cost of jockey pulley failures over the last 12 months amounts to £19,044.*

From the 'desired state' and the 'as is' situations, a statement of the problem was created, describing why the project should be carried out, as follows:

> *The elastic belt should have sufficient tension to ensure that it is suitable for assembly and that it maintains enough tension throughout the product life so that no slippage occurs, when subjected to a full load which is 100% wet. All this must be achieved at a reduced cost and with no adverse effects to the current product.*

Define the advocacy team

The reduction of SCR by introducing a new technology requires a cross-functional view of all parties involved. Therefore, the people who can provide process knowledge, expert knowledge and Six Sigma expertise, and those who will be relied upon to keep the problem fixed formed an advocacy team. Each member of the team has different roles and responsibilities. The Black Belt, for example, was the project leader and his responsibilities were to analyze data and lead the team. Production engineers and service technician were responsible for data collection and cascading down service issues. Several team members were also part of this team including suppliers, purchasing, development, health and safety and finance departments. A financial analyst was responsible for quantifying the potential financial benefits of the project. This includes the development of measurable and quantifiable financial metrics which reflect the potential savings of the project (Antony, 2004).

Potential financial benefits

There are two categories of Six Sigma project savings: 'hard' or bottom-line savings and 'soft' savings (Snee and Rodenbaugh, 2002). Hard savings are usually calculated by doing a financial analysis of year-to-year spending and trying to reduce spending and budget variances. Examples of hard savings include cost reduction and revenue enhancement. On the other hand, soft savings involve the reduction of cash tied up in inventory or the decreasing spending of capital, which are usually more difficult to quantify. Accordingly, the potential hard savings of the project are expected to come from a reduction in labor costs, parts used and parts failures. To estimate these hard savings, the company employs the budget sales numbers to calculate the potential number of products affected. In addition, reliability yields are calculated from the month

Table 4.1 Potential financial benefits

	Volume	Material savings per product	Labor savings per product	Total savings	Value of reliability improvement	Total annual savings
Product A	65,434	£0.44	£0.05	£0.49	£2784	£34,846.66
Product B	53,708	£0.67	£0.10	£0.77	£2156	£43,511.16
Product C	82,980	£0.44	£0.05	£0.49	£3530	£44,190.20
Product D	262,098	£0.67	£0.10	£0.77	£10,574	£212,389.46
Total					£19,044.00	£334,937.48

following the project closure. This will enable the yield in the financial year to be varied depending on when the project is completed. Table 4.1 shows the estimated savings of the project annually for the products affected.

Material costs are reduced by eliminating the number of parts used (7 parts). Labor savings are possible due to the elimination of the sub-assembly of jockey wheel and bracket unit, the line assembly of jockey wheel and bracket unit to base, and the fitting of the belt (Figure 4.2). Service savings consists of a reduction in SCR for the eliminated parts, that is jockey wheel and bracket unit, and a possible reduction in SCR for belts. In addition, there are some soft benefits such as a reduction in stock held in stores which were not quantified.

Project charter

To summarize the define stage and cascade down the project description, goals and benefits a project charter was carried out as shown in Table 4.2. It is a formal document that defines and describes the project at a high level and functions as the formal agreement for the project.

Due to subsequent Six Sigma process steps of the DMAIC methodology are built upon work completed during the define phase, the Six Sigma team ensured that the following deliveries were achieved before proceeding to the next phase:

- process linked to strategic business requirements;
- customer and CTQs characteristics identified;
- written team charter including rationale of the project, preliminary problem statement, scope, goals, milestones and roles and responsibilities;
- financial benefits identified and calculated.

The Six Sigma team agreed to the achievement of the above deliveries and proceeded to the measure phase.

4.3.2 Measure phase

The purpose of the measure phase is to establish techniques for collecting data about current performance that highlight project opportunities and provide a

Table 4.2 Project charter

Element	Description	Project charter
Problem statement	Describe why the project needs to be done	The belt should have sufficient tension to ensure that it is suitable for assembly and that it maintains enough tension throughout the product life so that no slippage occurs when subjected to a full load which is 100% wet. All this must be achieved at a reduced cost and with no adverse effects to the current product
Linkage to business	Identify linkage to business plan/department objectives	Reduction of SCR costs
Defect definition/ CTQ (measurable)	Define what is a defect and how it will be measured	Sufficient tension to ensure that it is suitable for assembly and maintains enough tension throughout the product life so that no slippage occurs when subjected to a full load. Tension is measured in Hertz
Definition of project scope	Set the boundaries of what is included and what is excluded	5 kg tumble dryers only
Project deliverables/ objectives (cost/ quality/timing)	State how the project will be assessed to determine its success and its specific deliverables (project outputs)	• Quality benefits: SCR reduction • Time benefits: Process cycle time reduction • Cost benefits: Material costs reduction less labor required for production parts standardization
Core team members	List the members who carry out the projects and the people who assist in the project execution	• Black Belt • Process owners (production engineers and service technician) • Team members (suppliers, purchasing, development, health and safety and finance)

structure for monitoring subsequent improvements. The following sub-steps were performed during this phase of the DMAIC methodology:

1. process mapping;
2. identify potential X's;
3. identify stakeholders' issues;
4. measurement system analysis (MSA).

Process mapping

Before collecting the data, the advocacy team decided to define the process under study. A process flow diagram was used as a visual representation of all major steps and decision points in the process. This helped in identifying potential

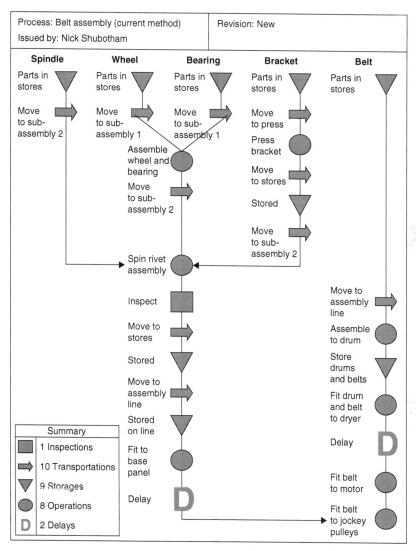

Figure 4.2 Process mapping of current method.

problem areas, understanding the process, giving directions for process improvements, and examining parts and information flow. For these characteristics the process flow diagram is considered as a key tool in identifying opportunities for improvement.

As it can be seen from Figure 4.2, the current process involves the assembly of five different parts. On the other hand, by introducing elastic belts seven parts can be eliminated and their associated number of non-value-added activities such as inspections, transportations, delays and storages. However, the Six Sigma team needed to identify potential factors affecting the tension on elastic belts.

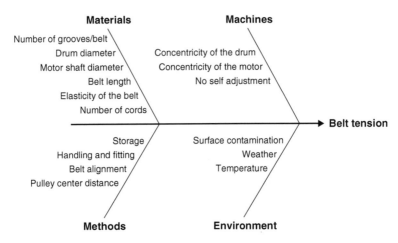

Figure 4.3 Cause and effect diagram.

Identify potential causal factors (X's)

The advocacy team identified potential X's during a brainstorming session. This session allowed the generation of a high volume of ideas rapidly which were then grouped using a cause and effect diagram or 'fishbone,' shown in Figure 4.3. The cause and effect diagram was later in the DMAIC methodology used for process capability studies.

Identify stakeholders' issues

Past attempts of introducing the elastic belts have failed in providing sufficient tension or offering too much tension which makes it difficult to fit the belt. The advocacy team decided to carry out a failure mode and effect analysis (FMEA) to overcome these difficulties. The objective was to discover and to correct the potential problems during the earliest stages of the project and to find out potential actions to prevent them. Table 4.3 shows the FMEA for the elastic belt. Note that FMEA states the potential failure modes or the manner in which the elastic belt can fail in conforming to the design requirement/customer expectations. In this case, the main concerns are breaking the elastic belt, belt migration and wearing. For each failure mode a potential effect of the failure was identified. The potential effects of the failure as perceived by the customer include no drum drive, overheating of load and slippage. For each effect a severity was assigned. A severity of '8,' in Table 4.3, indicates the potential of very high customer dissatisfaction due to a major disruption of the product performance. The potential causes in FMEA are indications of design weaknesses whose consequences are the failure modes. The occurrence score for each cause was then computed based on group expertise. For each potential cause the current controls to detect and prevent the failure modes were identified and its detectability assessed. Finally, a risk priority number (RPN) was estimated as

Table 4.3 Failure mode and effect analysis

Part name and function	Potential failure mode	Potential effects of the failure	Severity	Potential causes/mechanisms of failure	Occurrence	Design verification controls	Detectability	RPN
Belt To transmit drive from the motor to the drum	Breaking	No drum drive	8	Over tension	6	Supplier approval of application	9	432
	Breaking	No drum drive	8	Over tension	6	Supplier capability, service quality audit (SQA) visit	6	288
Must rotate without slipping 10 kg	Breaking	No drum drive	8	Over tension	6	Product tolerance capability	5	240
	Breaking	No drum drive	8	Over tension	6	Supplier analytical design	4	192
	Breaking	No drum drive	8	Over loading drum (torque)	2	Standards investigation into maximum load	10	160
	Breaking	No drum drive	8	Drum stalled	2	Drum over load testing	9	144
	Breaking	No drum drive	8	Damaged belt	2	Supplier recommended assembly techniques	9	144
	Breaking	No drum drive	8	Oil contamination	3	SQA visit	6	144
	Breaking	No drum drive	8	Drum stalled	2	Bearing testing under tension (limits established)	7	112
	Breaking	No drum drive	8	Drum stalled	2	Seal capability	7	112
	Breaking	No drum drive	8	Drum stalled	2	Seal location	7	112
	Breaking	No drum drive	8	Damaged belt	2	Supplier recommended storage conditions and shelf life	7	112
	Breaking	No drum drive	8	Overload drum	2	Test to failure or 10 k h	7	112
	Jump off motor shaft (belt migration)	No drive	8	Assembly	4	Spread of belt location on drum/shaft	8	256
	Wear	No drum drive	8	Motor lift (uneven wear)	10	Design tolerance/development testing	5	400
	Wear	No drum drive	8	Damaged grooves	5	Quality audit (QA)	4	160
	Wear	Slippage	5	Motor lift (uneven wear)	10	Design tolerance/development testing	5	250
	Wear	Slippage	5	Damaged belt grooves	5	SQA	4	100

the product of the severity, occurrence and detection to evaluate the risk level of the product and the process.

The failure modes with higher RPN should be solved first. Accordingly, the advocacy team decided to work with the supplier in order to develop a suitable elastic belt. Generally, designers of tumble dryers provide suppliers with the driver (motor) diameter, driven (drum) diameter, the center distance (rear panel) and the driven revolutions per minute (rpm). Based on these design parameters, suppliers design a suitable belt. However, variations on the actual drum diameter, the driver diameter and the center distance can lead to excessive or insufficient tension. Therefore, the project focused on identifying and quantifying variations on these design parameters to provide suppliers with the adequate information. Before evaluating the role of variations in these design parameters, the measurement system was analyzed.

Measurement system analysis

In the elastic belts, as in any process, there is variability. But there is also variability in the way the CTQs (i.e., tension, dimensions) are measured. The purpose of an MSA is to understand and quantify the different sources of variation within the measurement system. A measurement system consists of equipment, fixtures, gauges, instruments, mechanisms and appraisers that together make it possible to assign a number to a measured characteristic (Creveling et al., 2003). The variation in measurement can itself be divided in variation on repeatability and variation in reproducibility. Repeatability refers to the variation obtained when an operator uses the same gauge several times for measuring the identical characteristic on the same part. Whereas reproducibility refers to the variation obtained when several operators use the same gauge for measuring the identical characteristic on the same part (Antony, 2003). There are different methods to quantify repeatability and reproducibility (R&R) such as gauge capability R&R study and graphical (control chart) method. Gauge R&R studies quantify repeatability, reproducibility, total measurement variation and determine their contribution to the total process variation or specification. An initial step for developing a measuring system analysis using gauge R&R is to define the process to measure. In this case the team was primarily interested on analyzing the measuring system used to evaluate the dimensions of the motor, drum and rear panel. A useful criterion to evaluate the measuring system is the contribution of the total gauge variability to the total variability. It is said that a measurement system is capable if the ratio is lower than 10%. From Table 4.4, it can be seen that for the drum and rear panel components the total gauge R&R can be categorized as adequate showing a relative good measurement system capability. Whereas as for the motor it can be categorized as marginal since its ratio is between 10% and 30%. Measuring systems with less than 30% ratio are considered inadequate.

Having validated the measuring system, mapped the current process and identified the stakeholders' issues using FMEA, the advocacy team proceed to the analysis phase.

Table 4.4 Gauge R&R results

	Motor	Drum	Rear panel
Total gauge R&R	15.43	4.42	18.94
Repeatability	7.45	0.76	4.37
Reproducibility	7.98	3.66	14.57
Operator	1.23	0.00	1.84
Operator to part interaction	6.75	3.66	12.73
Part-to-part	84.57	95.58	81.06
Total variation	100.00	100.00	100.00

4.3.3 Analyze phase

The analysis phase allows the Six Sigma team to further target improvement opportunities by focusing on distinct project issues and opportunities. During previous phases, the Six Sigma team narrowed its focus on the belts' length which in function consists of three parts: drum, rear panel and pulley or motor. Consequently, most of the analysis phase was carried out to determine the probability of over or insufficient tension based on the actual variation of these components dimensions. Identifying how variations in parts affect the system help to (Creveling *et al.*, 2003):

- improve an operation by changing the mean values of the X's which have the most impact on the response;
- identify the factors whose variability has the highest influence on the response and target their improvement;
- identify the factors which have low influence and can be allowed to vary over a wider range;
- visualize the response, given a set of constraints.

As the tension of the belt can be affected due to variability in drum, base and pulley, the team calculated the capability of each individual part and then estimated the probability of failure in providing a required tension. From the drawing dimensions and tolerances the specification limits for the three CTQs were identified. Thus, according to the drawings, dimensions the drum diameter should be 568.5 ± 0.4 mm; the center distance 362.8 ± 0.4 mm (determined by the rear panel); and the pulley in the motor diameter 9 ± 0.1 mm. Having identified the upper and lower specification limits of the CTQ components, the process capability in the short and long term estimated by the sigma level was predicted. According to the sigma metric, processes or products may shift and drift in the long term. These shifts and drifts are caused by common changes in process and products which are generally not detected. However, data can be collected in 'rational subgroups' so that short-term variability can be assessed, coupled with long-term variability and shifts and drifts of the mean. The 'rational sub-grouping' technique quantifies the subgroup-to-subgroup variation by segregating product/process from different streams which generally generate drifts and shifts (e.g., suppliers, batches). Subgroups are selected

Table 4.5 Components capability

Component	Specification limits (mm)	σ level		Standard deviation		Mean Long term	DPMO	
		Short term	Long term	Short term	Long term		Short term	Long term
Rear panel	362.8 ± 0.4	5.88	−1.01	0.068	0.079	362.32	0.0042	843,752
Drum	568.5 ± 0.4	5.12	2.47	0.078	0.103	568.645	0.3056	6755
Motor	9 ± 0.1	4.29	3.37	0.0233	0.0267	9.01	17.86	395

in a way that appears likely to give the maximum chance for the measurements within subgroups to be alike and the maximum chance between subgroups to differ one from the other. For example, three different subgroups consisting of different batches of drums were 'rationally' selected. From them, 10 consecutive samples from a production line were taken consecutively to reduce dispersion on the subgroup. Thus, the short-term capability was calculated taking into account the variability within each batch, but not the variability between different batches, and assuming that the process/product is centered across the target. On the other hand, long term considers the total variability and does not assume that the process/product is centered.

From Table 4.5 it can be seen that the components have an adequate sigma level in the short term. For example, the rear panel has a short-term capability of nearly Six Sigma (i.e., 5.88σ). However, in the long term its capability is even negative (i.e., -1.01σ). This indicates that the mean of the process is felt outside the specifications limits. In addition, the difference between the long-term and short-term standard deviations is not a major contribution for the poor performance in the long term. The advocacy team attributed a poor sigma level long term to dimension errors, that is drawing specification vs. actual dimensions. These inconsistencies led to a reduction of the expected length needed for the desired tension. Based on this analysis the belts were re-specified by equaling the target specification value to the mean in the long term. However, for other cases it is compulsory to investigate the causes of drifts and shifts especially if these are greater than the average 1.5σ shift.

The tension of the belt is made up of the capability of parts above estimated. From the component tolerances of each component the geometric relationship between the components as they fit together to make an assembly were estimated. Using algebra and geometry the belt length required is 1855.49 mm (based on the nominal component values). Accordingly, not all parts have the same contribution or sensitivity to belt tension. For example, a unit change in magnitude of the pulley affects little to the final tension. On the other hand, changes in the drum diameter affect to a higher degree the response. The worst-case tolerance equation was also used to determine how a calculated assembly tolerance range is functionally related to the required belt length tolerance range, producing upper and lower specification limits for the belt length. The advocacy team provided to the supplier with the capability study above described to obtain a suitable elastic belt for this application.

As an outcome of the analysis phase, the Six Sigma team members had a strong understanding of the factors impacting the belt tension, including (Stamatis, 2003):

- key process input variables (the vital few 'X's that impact the 'Y');
- sources of variation (i.e., where the greatest degree of variation exists).

4.3.4 Improve phase

The improve phase has the objectives of considering the causes found in the analysis phase and selecting the target solutions. At the end of the phase, the causes found should be eliminated and the improvement committed achieved. Having specified the belt length required, the supplier provided a belt suitable for the application. The advocacy team decided to trial the supplied belts in a production line. This allowed the development of a fitting technique for operators and increasing confidence in the values developed. During the trial, belt tension was measured to assess the capability of the new technology.

From the capability study in Figure 4.4, it can be seen that the process is not centered. Although during the trial the entire sample was inside the specifications limits, it is expected that due to variation and the fact the process is not centered, around 983 Defects per Million Opportunities (DPMO) are expected. However, even better results can be achieved if the tension is increased to center the process. The team increased the belt tension and proceeded to confirm the new capability in a new production trial.

From the second process capability analysis in Figure 4.5, it can be seen that during the second trial the process is better centered and variation is slightly reduced. During this trial all the samples fell inside the specification limits, however, due to the inherited variation it is expected 13.8 DPMO. To prove that

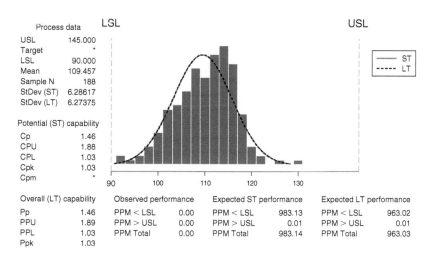

Figure 4.4 Process capability for the first trial.

the process has effectively moved towards a better performance after the belt tension was readjusted, a hypothesis test was carried out. From Figure 4.6 and confirmed by the hypothesis test it can be assumed that there is a statistical difference in the process capability improvement.

Tension is not the only CTQs. Additional tests were carried out to be sure that no adverse effects on other CTQs characteristics were present. This included tests for extreme operational conditions, safety assessments, post-life tests, etc. (see Table 4.6). As an outcome of the improvement phase the advocacy team validated and implemented the improved alternative.

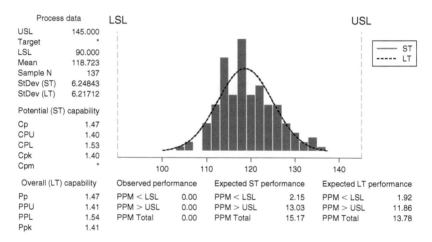

Figure 4.5 Process capability analysis for second trial.

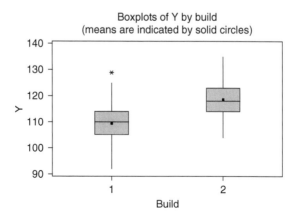

Figure 4.6 Box plot.

Table 4.6 Revised FMEA

Recommended actions	Responsibility and completion date	Actions taken	Severity	Occurrence	Detectability	RPN
Supplier to sign off application and insulation	Complete	E-mail confirming application	8	2	2	32
Review supplier history – SQA visit	Complete	CTQ data provided	4	2	2	16
Process capability study	Complete	Capabilities studies completed	4	2	2	16
Review actual tensions vs. predicted tensions. Cover full range of part sizes	Complete	Suppliers calculated tensions compared to actual	8	3	2	48
Establish max load drum can with stand	Complete	Post-life test with 12 kg point load	4	2	2	16
Low-voltage start up test	Complete	Test completed	4	2	2	16
Supplier to sign off application and insulation	Complete	E-mail confirming application	8	2	2	32
Run m/c with oily shaft	Complete	Machine starts ok	4	2	2	16
Review bearings post-life test for both systems	Complete	Front and rear bearings compared – old vs. new system	8	2	4	64
Review airflow pre- and post-life test	Complete	Airflow measurements taken post-life test compared to baseline	8	2	3	48
See above	Complete	Reviewed performance plus visual inspection	8	2	3	48
Supplier to provide information	Complete	Agreed method of packing/storage agreed	4	2	2	16
Supplier to sign off application and insulation	Complete	E-mail confirming application	4	2	2	16
SQA visit	Complete					0
Determine extent of misalignment	Complete	Belt difficult to fit if misaligned	8	2	3	48
Review belts post-life test	Complete	Analysis of belt post-life test	8	2	3	48
Process data/life testing	Complete	Audit check on initial run	8	2	3	48
Review belts post-life test	Complete	Analysis of belt post-life test	8	2	3	48
Process data	Complete	Audit check on initial run	8	2	3	48
None	Complete	Analysis of belt post-life test	8	2	1	16

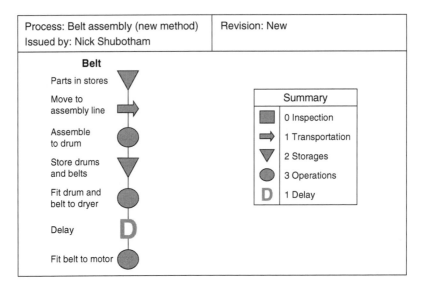

Figure 4.7 Process mapping for new method.

4.3.5 Control phase

The final step of the DMAIC methodology is to institutionalize product/process improvement and to monitor ongoing measures and actions to sustain improvement. Accordingly, the key actions taken during the control phase were:
1. a final confirmation of the process capability;
2. stockholder's issues revisited;
3. agree line inspection frequency.

Final confirmation of process capability

As a final confirmation of the process capability the proposed method was baselined using rational sub-grouping. In this instance, subgroups were selected using different tumble dryers manufactured in different production lines. Accordingly, the sigma level short term was 4.33 (7 DPMO) and the long term 4.19 (14 DPMO). The sigma shift in this case, which measures how well the process is controlled over time, was 0.14σ. Thus, it can be said that the introduction of elastic belts will improve reliability and nearly achieve the 3.4 DPMO Six Sigma goal. In addition, the targeted savings of labor and materials were achieved producing £335,000 in annual savings. These savings were possible, thanks to a dramatic reduction of process steps. Comparing the process map in Figure 4.2 with the new process in Figure 4.7, it can be noticed that the number of parts and assemblies were eliminated.

Stockholder's issues revisited

FMEA was carried out during the define phase to incorporate the stockholders' issues and avoid past failures when introducing elastic belts. An important part

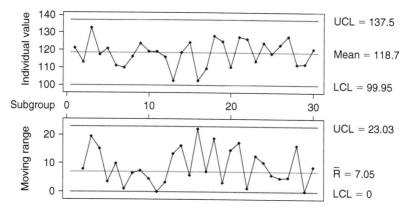

Figure 4.8 Control chart (I and MR) for belt tension.

of FMEA is to recommend actions and implement them. In this case, the Black Belt was responsible for implementing a follow-up program to ensure that all recommended actions are being implemented or adequately addressed. After corrective actions were implemented, the RPN for the corrected product/process was re-calculated.

As can be seen from Table 4.6, the RPN decreased as a result of the actions taken. Using FMEA, several issues identified by the stakeholders were addressed to facilitate the introduction of the new technology.

Agree line inspection frequency

The advocacy team developed a monitoring plan for the belt tension that requires periodical sampling. The purpose of the sampling is to detect any deterioration of the belt tension performance. During this process monitoring, tension is recorded during sampling and plotted on a control chart. The advocacy team used individuals (I) chart and moving range (MR) control chart to highlight conformance to the specification of the elastic belts. Considering both charts together allows for monitoring of both the tension level and tension variation, as well as for detecting the presence of special causes (see Figure 4.8).

As the final step of the DMAIC methodology, the project was documented and the lesson learned transferred to similar products for replication. This project was completed and formally closed 6 months after the initialization date.

4.4 Conclusion

The company where this case took place launched Six Sigma as a catalyst for focusing on customer satisfaction, to accelerate performance and to leverage best practices from the industry. The project shown here is an example of the companys' Six Sigma implementation. This project was targeted to customer satisfaction by improving the reliability of the product. In addition, by improving reliability it was also possible to reduce the number of parts, reduce labor and reduce the amount of money tied in inventory. To achieve these results using Six Sigma DMAIC methodology, several statistical tools and techniques were employed. It is important to mention that the tools used on each phase of the methodology can also find application in a different phase. There are no fixed rules that specify the correct order of tool usage; each project can evolve in its own unique manner, demanding its own order or tool usage. The role of the Six Sigma practitioner is to determine the tool selection and their use (Breyfogle *et al.*, 2001). The savings generated from this project were estimated at around £335,000 per annum. Several projects in the company show similar financial benefits, leading the company to implement Six Sigma company-wide including non-manufacturing operations such as finance, marketing, sales, accounting and human resources. In addition, design for Six Sigma has also been implemented as an effective way to prevent defects and incorporate the voice of the customer from the design stage.

Additional to the final benefits achieved, several further aspects of the project were deemed worthy of notice. Teamwork was a fundamental element within this project. The value of teamwork formed by cross-functional members launched a sense of ownership, better communication, team working and overall view of the organization. Suppliers were also part of the Six Sigma team. Working with them allowed the company to obtain a deeper knowledge of the process inputs. From the suppliers' point of view, their participation of the project encouraged the understanding of their customers' needs. Using Six Sigma the team employed several statistical tools to deal with variability and how to collect and use data so that informed decisions were made. With this new mindset, known as statistical thinking, the team understood that all work occurs in a system of interconnected processes and variations exist in all processes. The understanding and reduction of variations were the key to success.

Acknowledgment

Special thanks to Ghislaine van der Burgt for her assistance in proof reading the final version of the manuscript.

References

Antony, J. (2003). *Design of Experiments for Engineers and Scientists.* Oxford: Butterworth Heinemann.

Antony, J. (2004). Some Pros and Cons of Six Sigma: An Academic Perspective. *The TQM Magazine*, 16(4), 303–306.

Breyfogle III, F.W., Cupello, J.M. and Meadows, B. (2001). Managing Six Sigma: A Practical Guide to Understanding, Assessing and Implementing the Strategy that Yield Bottom-line Success. NY, USA: Wiley.

Creveling, C., Slutsky, J. and Antis, D. (2003). *Design for Six Sigma in Technology and Product Development.* New Jersey: Prentice Hall.

Snee, R. and Rodenbaugh, W. (2002). The Project Selection Process. *Quality Progress*, 35(9), 78–81.

Stamatis, D. (2003). Six Sigma for financial professionals. Hoboken, NJ: John Wiley and Sons Inc.

5

An application of Six Sigma in an automotive company

Jiju Antony, Maneesh Kumar and M. K. Tiwari

5.1 Introduction

This chapter deals with a case study performed in an automobile company to eliminate the defects in casting product using Six Sigma problem-solving methodology. The application of Six Sigma methodology (DMAIC) reduced the number of defects in the engine manufacturing process and thereby improving customer satisfaction and business profitability for the company. The five-step Six Sigma methodology (DMAIC) used for this case study is depicted in Figure 5.1.

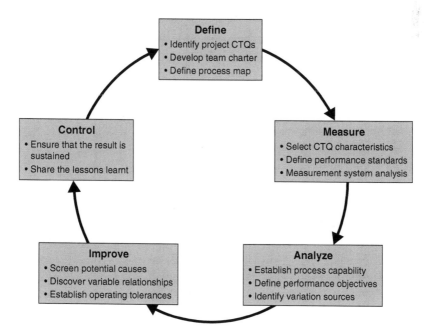

Figure 5.1 The five- step methodology of Six Sigma.

The five-step process of the case study started with the define phase. The *Define phase* involved identifying a project's critical to quality (CTQs) characteristic driven by the voice of customer (VOC), followed by developing team charter, and finally defining a high level process map connecting the customer to the process and identifying the key inputs and requirements.

In the *measurement phase*, the team identified the key internal processes that influence CTQs and measures the defects currently generated relative to those processes. The project's CTQs were selected with the help of a fishbone diagram, Cause and Effect matrix, Pareto chart, etc. A Gauge R&R (repeatability and reproducibility) study was conducted to check whether the measurement system is acceptable or not.

The *Analyze phase* mainly consisted of three steps: establishing process capability with the help of capability indices, defining performance objectives by the team benchmarking, and identifying the sources of variation by performing an ANOVA test and hypothesis testing. Based on the above information, root causes of defects and their impact on the business/process were identified.

The *Improvement phase* helped the team to confirm the key variables and quantifies their effects on the CTQs. Design of Experiment (DoE) was performed to optimize the process. By implementing the optimum parameters into practices, the performance can be improved and financial goal is reached.

In the last step of the DMAIC process, i.e. the *Control phase*, the improved data of significant factors, which were identified from the experimental design, were monitored and whole process were well-documented to ensure that the improvements were sustained beyond the completion of project.

The remainder of this chapter is arranged in the following way: In section 5.2, the introduction to the case study is presented and the solution to the problem is explained in terms of the five steps of the DMAIC approach. Section 5.3 illustrates the key results and financial savings generated from the project. This is followed by a discussion on the managerial implications in Section 5.4. Teaching notes on the key ingredients necessary for the implementation of Six Sigma projects is presented in Section 5.5. Finally, Section 5.6 presents the concluding remarks and the significance of the project.

5.2 Case study

This case study deals with the reduction of casting defects in an automobile engine. This problem was tackled using a Six Sigma DMAIC problem-solving methodology. The basic equation of Six Sigma $Y = f(x)$ defines the relationship between a dependent variable 'Y' or outcome of a process and a set of independent variables or possible causes which affect the outcome. In the present case study, 'Y' is the high customer dissatisfaction due to an unacceptable number of casting defects in the engine. Six Sigma problem-solving methodology (DMAIC) is recommended when the cause of the problem is unclear (Snee and Hoerl, 2003). This project was of the highest priority to senior managers within the company as it was known that an effective solution to this problem will have a significant impact on the bottom-line. Moreover, it was clear to the team members as well as project champion that elimination of this problem will have colossal impact on customer satisfaction.

Customers repeatedly reported the high casting defect rate in the engine over a period of time, which was unacceptable to the manufacturer. This led to immense customer dissatisfaction, and a threat to the reputation of the company. The root causes of this defect were sand fusion and metal penetration during casting of the product. These two causes were responsible for the higher casting defect in the engine. Sand Fusion is a defect encountered by the foundry shop when the molten metal, due to its high temperature, fuses with the sand of mold and core. Another reason for sand fusion may be the low thermal strength of the sand used for the process of making cores. Metal penetration is due to porosity remaining in the cores used for manufacturing of the casting product. In thin sections, slight porosity may result in the considerable weakening of the cores and thus the metal may penetrate and jam the mold cavity. Defect of metal penetration occurs in various locations in the casting product due to sand fusion.

5.2.1 Define phase

This phase of the DMAIC methodology was to define the scope and goals of the improvement project in terms of customer requirements; and to develop a process that delivered these requirements. The first task was to develop a project charter to help team members clearly understand the scope and boundaries of the project: objectives, duration, resources, roles of team members, estimated financial gains from the project, etc. This created a sense of ownership for the project. It also prevented delivery of mixed messages between project managers and team members. The project team members included a champion, black belt, process owner, and a green belt.

In the define phase of the methodology, there were many questions asked by the team members during the project charting session:

- What is wrong in the production of the casting?
- Where is the problem?
- How big is the problem?
- What is the impact of the problem?

The Six Sigma team ensured that the following points had been looked into prior to embarking on the measure phase.

The goal statement of the project defined by the team members was the reduction of defects per unit (DPU) that should result in an immense reduction in the cost of poor quality (COPQ). The team conducted destructive testing on the casting produced on different dates to identify the root causes of the problem. After performing a number of brainstorming exercises and using a multi-voting method, the team members arrived at the conclusion that the defect in the casting was due to sand fusion and metal penetration during casting of the product. The team also concluded that the root cause of sand fusion and metal

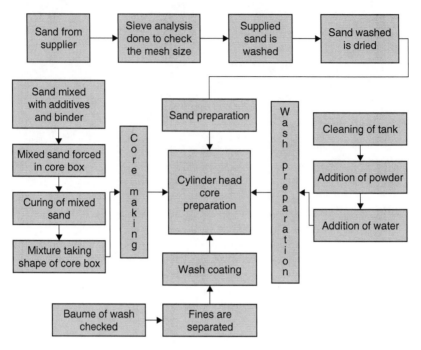

Figure 5.2 Process mapping for core preparation of casting.

penetration was the porous core. The impact of the problem was very severe because it was a main cause of casting defect, which led to high warranty failures and customer dissatisfaction.

The team focused on the following processes for enhancing customer satisfaction and reducing COPQ in the foundry:

- Sand preparation
- Core making process
- Wash preparation and coating

These processes were selected based on the sound engineering knowledge, expertise of team members with the process, as well as taking into account the steps used in making of casting.

5.2.2 Measure phase

This phase was concerned with selecting one or more product characteristics, mapping the respective process, making the necessary measurements, recording the results on process control cards and establishing a baseline of the process capability or process performance. Process mapping provideed a picture of the steps needed to create the output or process 'Y'. It was a pictorial representation of the process, which helped to identify all value and non-value added steps, key process inputs (Xs) and outputs (Ys). Figure 5.2 shows the

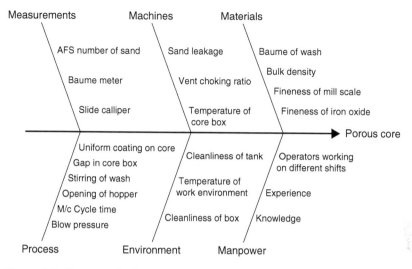

Figure 5.3 Cause and effect analysis of porous core.

process mapping of casting product with three core processes; sand preparation, core preparation and wash preparation and coating.

Having mapped the process, the team proceeded to analyze the potential causes of defects. Although this task was carried out in detail during the analyze phase of the Six Sigma methodology, the team had to plan about how well the 'Xs' are controlled in the given process. Figure 5.3 illustrates the cause and effect analysis for the problem encountered by the company. The effect in this example is the porous core. The output of the cause-and-effect diagram depends to a large extent on the quality and creativity of the brainstorming session.

The cause and effect analysis showed that the process variables affecting porous core were SL, BP, ANS, GCB, BD, and VCR (the full names of these variables cannot be revealed in the case study due to confidentiality agreement between authors and the company). It was also observed by the team members that the contribution of porous core in casting defect was over 80%. Having constructed the cause and effect diagram, the team then created a cause and effect matrix. Table 5.1 shows the cause and effect matrix showing the customers' needs and the relative importance of the process characteristics which are critical to customers.

The next step was to define performance standards according to customer requirements. The goal of performance standard is to translate the customer needs into measurable characteristics; based on the specification limits, performance standards for each process parameter are established.

Having established the key process parameters and the critical to quality characteristic (depth of porous core), it was essential to establish the accuracy of the measurement system and the quality of data. A data collection plan was established to focus on the project output and also to carry out the standard setting exercise for the same. A Gage R&R study was conducted to identify the sources

Table 5.1 Cause and effect matrix

Customer needs	Process characteristics					
	Depth of porous core	BW	PW	SW	SHW	Importance
Uniform coating	H	H	L	M	L	8
Proper thickness	H	H	L	L	M	9
Filling of porous core	H	M	L	L	L	9
Weighted requirements	130	112	26	42	44	

H Strong relationship = 5, M Medium relationship = 3, L Weak relationship = 1.

Table 5.2 Results of gage R&R study

Source	Variance	% Contribution of variance
Total gage R&R	1.62E − 03	6.08
Repeatability	1.60E − 03	6.02
Reproducibility	1.67E − 05	0.06
Part-to-part	2.50E − 02	93.92
Total variation	2.67E − 02	100.00

of variation in the measurement system and to determine whether the measurement system is capable or not. One may redesign the gauge and perform proper maintenance of instrument when repeatability is large compared to reproducibility. When reproducibility is large compared to repeatability, then the operator needs to be better trained in how to use and read the gauge instrument and a suitable fixture was needed to help the operators use the gauge more consistently. The measurement system is considered acceptable when the measurement system variability is less than 10% of total process variability (Antony *et al.*, 1999). However, the system may be acceptable when the measurement system variability is between 10% and 30%, and for above 30%, the measurement system is not considered acceptable. The study is performed to check the accuracy of gauges used for the measurement of characteristics as well as the accuracy of the worker in performing their operations on the machine. The gage R&R study performed on the system showed a variation of 6.08%, which implies that the measurement system is acceptable. Table 5.2 illustrates the results of the Gage R&R study carried out during the measurement phase of the project.

The baseline process capability (C_{pk}) was also established in this phase. The C_{pk} value based on the existing process conditions is estimated to be 0.49. This clearly indicates that process performance is poor and it clearly needs improvement.

5.2.3 Analysis phase

The first step in this phase was to gather data from the process in order to obtain a better picture of the depth of porous core values under different process

Table 5.3 Results of regression analysis for process parameters

Process parameters	P Value
SL	0.002*
ANS	0.414
BW	0.060
BP	0.155
BD	0.003*
FT	0.104
VCR	0.001*

Note: *indicates the significance of process parameters at 1% and 5% significance levels.

conditions. Data pertaining to factors affecting the response was collected over a period of 36 days from different shifts of the day.

In routine foundry production, the casting is shipped to the customer if the mechanical test data satisfies the requirement of the standard and no defect occurs in the casting. Data related to factors affecting the response (depth of porous core) is analyzed to determine not only the relationship between the process parameters and the response but also to determine the direction of process improvement.

In the analysis phase, it was important to identify the possible sources of variation which lead to the casting problem. Moreover it was also important to understand the causes for poor process capability. The aim of the project team was to enhance the process capability by reducing variation in the process.

At this point, it is imperative to identify the parameters that are significant to the process so that they can be brought under statistical control. The improvement goal of the project is defined statistically through benchmarking with an automobile company located in USA. A simple regression analysis is performed to determine the significance of process parameters. It is concluded from the regression analysis that the variables having a 'P' value of less than 0.05 and 0.01 are statistically significant for further study. Table 5.3 shows that SL, BD and VCR are the parameters that need to be further optimized and controlled. The optimization of these parameters will yield us an optimum response (i.e., depth of porous core).

5.2.4 Improve phase

In this phase, it was decided to perform a designed experiment with the above three process parameters (SL, BD and VCR) identified from the analysis phase.

Design of Experiments (DoE) was conducted using the three process parameters identified from the previous phase. Each process parameter was studied at two levels in order to keep the size of the experiment to a minimum as well as due to time and cost constraints. A coded design matrix for the three significant variables (SL, BD and VCR) is depicted in Table 5.4. A 2^3 full factorial design was chosen so that both main effects and interaction effects among the parameters

Table 5.4 Coded design matrix for the process variables to conduct DoE

SL	BD	VCR
1	1	1
1	2	1
1	1	2
1	2	2
2	1	1
2	2	1
2	1	2
2	2	2

Table 5.5 Results of 2^3 full factorial experiment

SL	BD	VCR	Depth of porous core (in mm)		Average of depth of porous core
			Replication 1	Replication 2	
1	1	1	0.75	0.65	0.7
1	2	1	0.6	0.6	0.6
1	1	2	0.8	1.0	0.9
1	2	2	0.85	0.75	0.8
2	1	1	0.90	1.0	0.95
2	2	1	0.8	0.9	0.85
2	1	2	0.9	1.1	1.0
2	2	2	0.9	1.0	0.95

could be investigated. In order to have sufficient degrees of freedom for studying both main effects and interaction effects, each trial condition was replicated twice (Antony, 2003). Table 5.5 illustrates the results of the experiment with average depth of porous core as the response of interest. The average depth of porous core before the experiment was 1.25 mm.

As the objective of the experiment was to minimize the depth of porous core, the first objective of the analysis was to determine the effect of process parameters and to understand the presence of any interactions, if present. Figures 5.4 and 5.5 illustrate the main effects plot and the interactions plot, respectively. In order to determine the statistical significance of both main and interaction effects, it was decided to construct a Normal Probability Plot of effects (see Figure 5.6). Figure 5.6 indicates that only the main effects were statistically significant at a 10% significance level. None of the interactions were statistically significant although the interaction plot (Figure 5.5) suggesting slight interaction between VCR and SL. The main effects plot (refer to Figure 5.4) indicates that the optimum levels of process parameters for minimizing the depth of porous core are:

BD – High level

VCR – Low level

SL – Low level

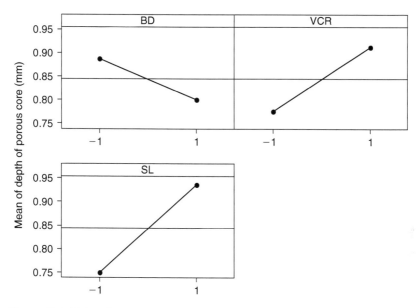

Figure 5.4 Main effects plot on depth of porous core.

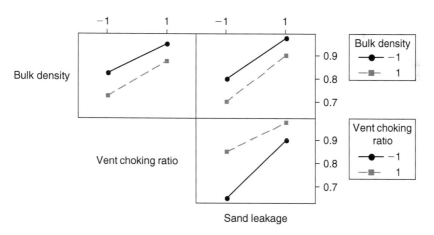

Figure 5.5 Interactions plot showing the interactions among the process parameters.

Confirmation trials were carried out using the optimal settings and the average depth of porous core was computed to be 0.80 mm. Moreover, the process variability was significantly reduced as well. The process capability (C_{pk}) had improved from 0.49 to 1.28. This clearly indicates a significant improvement of process performance.

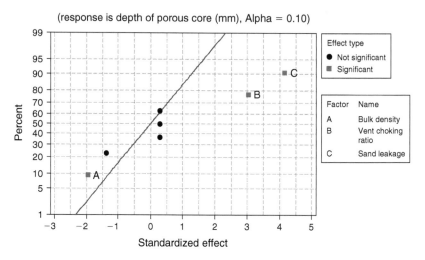

Figure 5.6   Normal probability of effects for the experiment.

5.2.5 Control phase

The real challenge of Six Sigma methodology is not in making improvement of the process but in keeping the optimized results sustained. This needs standardization and constant monitoring and control of the optimized process. An extensive training program for the process related personnel was conducted within the company where the case study was performed. It is well known that real improvement will only come from the shop floor. Process sheets and control charts were made so that the operator takes preventive action before the critical process parameters and critical performance characteristics go outside the control limits. A complete database was prepared to sustain the improved result. Proper monitoring of the process helped to detect and correct out-of-control signals before it resulted in customer dissatisfaction. Normal silica or Chromite sand was mixed with shell sand of a recommended AFS number to get best results. Implementation of aforementioned suggestions resulted in further improvement of process capability and process yield.

Run charts for the depth of porous core were constructed (Figures 5.7 and 5.8) before and after the improvements were made to the process. The purpose of the run charts here was to analyze variability in the porous core around its mean value. Figure 5.8 (after improvement phase) shows that all the points are within the specifications and variability in porous core has been reduced significantly.

5.3 Key results and financial savings generated from the project

The production department manufactured 100,000 units of casting to cater to the needs of the customers. Production was carried out in two shifts per day,

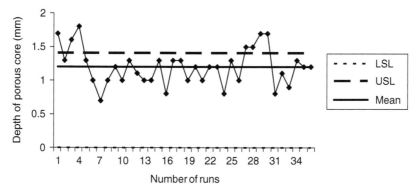

Figure 5.7 Run chart for porous core before improvement.

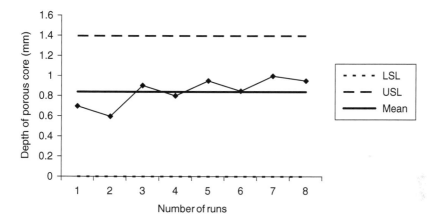

Figure 5.8 Run chart for depth of porous core after improvement.

with 8 hours per shift, and there were six working days per week. The casting manufacturing involved a series of sequential steps that are as follows:

a) Making of core in a core shooter machine (P_1)
b) Finishing of core (P_2)
c) Coating of finished core with a wash (P_3)
d) Drying the wash coating by passing through heating section (P_4)
e) Assembly of core in an automated casting section for metal casting (P_5)
f) Removing the unwanted part from casting by passing through fettling section (P_6)
g) Cleaning and polishing of casting (P_7), and
h) Finally, the quality of casting is tested in Inspection/testing unit (P_8).

After achieving the optimal condition and keeping the result sustained, the team carried out a cost benefit analysis of the whole project. The cost incurred by the company in manufacturing of the casting product was divided into four categories: labor cost, raw material cost, operating expenses, and other overhead

Table 5.6 **Total cost (excluding raw material cost) incurred in manufacturing a unit product**

Manufacturing steps	Labor cost/ unit	Operating cost/ unit	Overhead cost/ unit	Total cost
Process 1 (P_1)	0.02	0.24	0.07	
Process 2 (P_2)	0.03	0.29	0.10	**7.77**
Process 3 (P_3)	0.04	0.37	0.12	
Process 4 (P_4)		0.49	0.14	
Process 5 (P_5)	0.06	0.77	0.24	
Process 6 (P_6)	0.06	0.95	0.30	
Process 7 (P_7)	0.08	1.20	0.36	
Process 8 (P_8)	0.10	1.33	0.41	
	0.39	**5.64**	**1.74**	

All cost are expressed in US dollars.

cost. All these costs were incurred in manufacturing the product passing through various processing stages. A break-up of all these costs in manufacturing a single unit that passes through eight sections is listed in Table 5.6.

As the raw material is processed through various machines, some value is added at every stage of production. Therefore, a defect coming at any stage of production is considered as a loss to the organization. However, the impact of defects on the organization increases with an increase in the stage at which they are identified, because at each stage some value is added to the product and each value adding activity has associated cost. The impact of defects in the form of delay reschedule, rework, delivery delay and more inspection also increases with the increase in the number of stages at which they are identified.

Savings generated from a Six Sigma project is divided into two parts: firstly, savings spawned from raw material and secondly from the other sources that includes labor cost, operating cost, and other overhead cost. The details of cost saving from raw materials are listed below. They demonstrate the power of Six Sigma as a strategy for achieving significant financial savings to the bottom-line of the organization. This would have also assisted the senior managers of the company to appreciate the power of Six Sigma.

5.3.1 Details of cost saving generated from raw material

Casting Required – 100,000/year
Core Required – 100,000/year
Core rejection before Project – 0.194 DPU
Core rejection after Project – 0.029 DPU
So core rejection reduced by 85%
85,000 number of cores saved per year
Shell sand required for 85,000 cores = 336,000 kg
Shell sand cost per kilogram = $0.30/kg
Total Cost Saved = $100,053.33 US

Table 5.7 gives details of savings generated from other sources which included labor costs, operating costs, and other overheads by the organization when the

Table 5.7 Break up of cost saving at each stage of production

Manufacturing steps	Labor cost/unit	Operating cost/unit	Overhead cost/unit	Total cost/unit	% Defect before Six Sigma	% Defect after Six Sigma	Annual impact before Six Sigma	Annual impact after Six Sigma	Savings after implementation
Core shooter machine	0.02	0.24	0.07	0.33	7.0	0.029	2310	9.57	2300.43
Surface finishing of core	0.03	0.29	0.10	0.42	0.2	0.012	84	5.04	78.96
Wash coating	0.04	0.37	0.12	0.53	0.1	0.007	53	3.71	49.29
Heating section		0.49	0.14	0.63	3.0	0.008	1890	5.04	1884.96
Casting unit	0.06	0.77	0.24	1.07	4.0	0.010	4280	10.70	4269.30
Fettling unit	0.06	0.95	0.30	1.31	0.7	0.0087	917	11.40	905.60
Cleaning & polishing	0.08	1.20	0.36	1.64	1.0	0.0031	1640	5.08	1634.92
Inspection/testing unit	0.10	1.33	0.41	1.84	0.2	0.0062	368	11.41	356.59

Total savings = $ 11480.05 US

* All cost are expressed in US dollars.

Table 5.8 Comparison before and after improvement based on key metrics

Key metrics used	Depth of porous core before improvement	Depth of porous core after improvement
Yield	82%	97.14%
Capability indices	0.49	1.28
Process mean	1.202 mm	0.843 mm
Process standard deviation	0.277 mm	0.137

defect was detected at a particular stage. The cost saving from the other sources comes out as $11,480.05 US.

It is deduced from Table 5.7 that the percentage defect coming in each stage has decreased considerably. Thus, there is an improvement of $111,533.38 of the total profit in monetary terms for the company after Six Sigma implementation. This profit in monetary terms is attributed to savings generated from raw material and other sources that included labor, operating, and other overhead costs incurred by organization.

The following are the equations used to calculate the Yield of casting product.

$$DPU = \text{no. of defects found/no. of units processed} \tag{1}$$
$$Yield = e^{-DPU} \tag{2}$$

Table 5.8 presents the key results of the study showing the key metrics used in the study. These metrics clearly indicate the performance improvements achieved by the process after implementation of Six Sigma methodology.

The data shown in Table 5.8 were collected over a period of 4 days from different shifts. It was observed that porosity in the core reduced drastically. Process capability of the system was increased from the previous value of 0.49 to 1.28 showing tremendous improvement in the production system.

5.4 Managerial implications

In this research work, the top level management of the organization realized the importance of strategic initiative needed for successful deployment of Six Sigma. A brainstorming session was conducted to contemplate on the changes in process, management practices, and environment changes required in securing improved quality of space work life and better business processed – both of which are requisites for customer success and, ultimately, achieving measurable and sustainable performance gains.

The next crucial step for the management was the selection of the right Six Sigma project so as to align the project with the strategic goals of the business. The top management called a meeting of senior managers from different production units to discuss the customer complaints experienced in their respective units. The purpose of the meeting was to sort out the problems that may be causing customer dissatisfaction. It was found that the majority of complaints coming from different parts of the country were related to casting defects in engines, which were endangering customer loyalty towards the company. This problem was encountered by customers across the country.

After achieving the strategic goals and keeping the result sustained, the management communicated the results and benefits generated to all its employees. Management felt that open communication and information sharing could promote a common culture and innovative behavior within the organization. The importance of adequate and focused (e.g., role related) training for successful implementation of Six Sigma cannot be overemphasized. Costs associated with training is perhaps viewed as the best investment opportunity by the management. The organization followed a structured methodology for managing change. The management was committed to making Six Sigma a top priority within their business environment. On the strength of this, the organization decided to provide coaching, counseling, and training to the people involved in the project; give rewards and share profits from the projects with its employees to motivate them to bring about a cultural change within the organization; recognize and reinforce desired improvement alternatives and desired behaviors, which included periodic project reviews between the Management and the people responsible for improvement activities.

5.5 Key ingredients of Six Sigma project implementation

For the successful implementation of Six Sigma projects, certain factors or ingredients should be in place. The success of Six Sigma is linked to a set of cross-functional metrics that lead to significant improvement in customer satisfaction and bottom-line benefits. The approach involves identifying projects that target the customer's concerns and have the potential for significant payback. This is accomplished by putting the quality and process improvement tools into the hands of a large number of managers, engineers and operators throughout the organization.

The application of Six Sigma requires top management involvement and provision of appropriate resources and training (Halliday, 2001). Senior Managers within the organization must be taught the principles of Six Sigma to enable the restructuring of the business organization and change their attitude towards a more disciplined approach. Continuous support and uncompromising commitment from top management are indispensable to the successful introduction and development of Six Sigma program (Pande et al., 2001). Adjustments to the culture of the organization and a change in the attitudes of the employees are again very important at the introduction stage of Six Sigma in any organization.

Employees have to be motivated and accept responsibility for the quality of their own work. Implementation of Six Sigma programs require the right mindset and attitude of the people working within the organization at all levels (Antony and Banuelas, 2001). The people within the company must be made aware of the changes. The result obtained by implementation of the program must be made public and the result should not only be related to success stories but also admit and communicate stumbling blocks. Doing such things will help other projects in the pipeline to avoid the same mistakes and learn from the mistakes.

There is also the need for an effective organizational infrastructure to be in place in order to support the Six Sigma introduction and development program within any organization. Extensive training must be provided to all professionals, such as project champions, black belts and master black belts so that a clear vision of how Six Sigma can help the organization meet its business goal is developed. The belt system ensures that everyone in the organization is speaking the same language so as to make the execution of project much easier throughout the organization.

A healthy portion of Six Sigma training involves learning the principles behind the methodology that takes the form of projects conducted in phases generally recognized as DMAIC (Define–Measure–Analyze–Improve–Control). It is not just the DMAIC methodology that makes the application of Six Sigma successful in organizations, rather it is the collection of tools and techniques which are integrated into DMAIC in a sequential and rigorous manner. Moreover, DMAIC creates a sense of urgency by emphasizing rapid project completion in 3 to 6 months.

Six Sigma builds on improvement methods that have been shown to be effective and integrates the human and process elements of improvement. The human elements of process improvement may include teamwork, customer focus and culture change. The process elements of process improvement include: understanding the types of process variation, process capability indices, process stability, and Design of Experiments (DoE) to reduce process variation and hence improve process performance.

Six Sigma is based on the scientific method which utilizes statistical thinking. Statistical thinking, therefore, is fundamental to the methodology of Six Sigma, which is action-oriented, focuses on processes and defect reduction through variation reduction and quality/process improvement goals. The core principles of statistical thinking, and the role of management in statistical thinking, for improving business performance will continue to grow in the coming years.

5.6 Conclusion

This chapter presents a case study of an automobile company, showing how the effective introduction and implementation of a Six Sigma program in organizations can lead to business profitability. The application of DMAIC methodology has been extremely valuable in reducing the casting defect. The yield of the process improved from 82% to over 95%. Moreover, the capability of the process has been significantly improved from 0.49 to 1.28. The estimated savings generated from this project was over $111,000 U.S. The project team members included a champion, black belt, process owner, and a green belt. This project was targeted to enhance customer satisfaction by reducing the casting defect in an automotive engine. The results of this project provided greater stimulus for the wider applications of Six Sigma methodology across the company in the future.

References

Snee, R.D. and Hoerl, R.W. (2003). *Leading Six Sigma – A Step by Step Guide based on Experience at GE and other Six Sigma companies*, FT Prentice-Hall, NJ, USA.

Pande, P., Neuman, R. and Cavanagh, R. (2001). *The Six Sigma Way*. McGraw-Hill, New York, NY.

Antony, J., Knowles, G. and Roberts, P. (1999). Gauge Capability Analysis: Classical versus ANOVA. *Journal of Quality Assurance*, 6(3), 173–182.

Antony, J. (2003). *Design of Experiments for Engineers and Scientists*. Butterworth Heinemann, Oxford, UK.

Halliday, S. (2001). So what exactly is Six Sigma?, *Works Management*, Vol. 15, No. 1, p. 15.

Antony, J. and Banuelas, R. (2001). A Strategy for Survival. *Manufacturing Engineer*, 80(3), 119–121.

6

An application of Six Sigma methodology to the manufacture of coal products

Edgardo J. Escalante Vázquez and
Ricardo A. Díaz Pérez

6.1 Introduction

This project is an application of the Six Sigma methodology in a company that manufactures coal products, with the general objective of improving a specific process and therefore to satisfy its customers' requirements. Although this case is not part of a company-wide effort to adopt Six Sigma, it was useful to show the upper management about its importance and what they may expect should they decide to embrace it. The specific objective of this investigation was to analyze the elements that constitute the Six Sigma methodology, and to apply it to improve a specific coal manufacture process where the highest amount of scrap is generated.

This study is divided into the following sections: company background, the application of the DMAIC (Define–Measure–Analyze–Implement–Control) methodology, the economic benefits obtained, the final conclusions, comments, and lessons learned from the case study.

6.2 Company background

The company where this study was undertaken is a multinational chemical company. Its Mexican operations are based on the manufacture of coal products for more than 40 years. Due to confidentiality, it is not possible to reveal the exact nature of their products.

Although the company does not have the ISO 9000 certification, its quality system is based on it. Furthermore, they use and apply lean manufacturing principles, and the theory of constraints too. The main processes are being monitored by statistical process control, and approximately 85% of the manufacturing factors are statistically stable. Additionally, 50% of them have a Cpk (real capability index) greater than or equal to 1.33 (Cpk = 1.5 for a Six Sigma process).

The Six Sigma initiative has been proposed to the upper management, but it has not been accepted due to current budget constraints. The company's main current difficulties are a high worker turnover, equipment failures, accidents, and high operating costs. There is only one green belt – trained on his own while working for this company – that was part of the project team too.

6.3 Define phase

In this phase the customers are defined along as their needs (CTXs). A project charter is developed as well as the business process map through an SIPOC (supplier–input–process–output–customer) diagram.

6.3.1 Product, customers, and critical-to-quality

The product is a D-by-D rectangular solid of variable length (see Figure 6.1). One set of customers was defined: external only. These customers receive the final product and they are interested in the *yield* they can achieve by using the product. The yield is related to product's consistency which in turn is dependent on the *opposing* or *resulting force* during the pressing or product extrusion operation in the forming process (see Figure 6.1 for a description of the above).

6.3.2 Project charter

A project charter is a description and a guide for the team members to be focused on the project, and to follow a logic sequence of steps to fulfill their objectives. It includes the business case, a statement of the problem, goals, roles and responsibilities, project scope, milestones, and a communications plan (Waddick, 2001).

6.3.3 Business case

During the past 2 years, raw materials (RM) efficiency has been low. Specifically, an increase of 166% in reprocessing of RM occurred during that period because

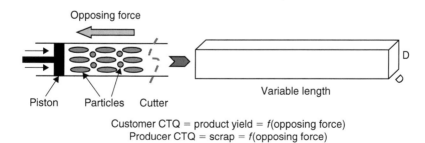

Customer CTQ = product yield = f(opposing force)
Producer CTQ = scrap = f(opposing force)

Figure 6.1 Product description and critical-to-quality (CTQ).

of the opposing force being outside of its specification limits. This high figure called the attention of the quality manager who decided to study the situation and to develop counter measures to it. The expected project benefits are the reduction of the manufacturing cost by an amount of 599,200 USD annually. This figure was obtained as follows: the year 2002 overall scrap level was 1.7%, costing 952,000 USD (cost of poor quality, COPQ). Now, based on a numeric goal of reducing the 1.7% scrap level to 0.63% (see Section 6.3.5), the estimated benefits will be $1.70 - 0.63 = 1.07\%$ or $0.0107(952,000)/0.017 = 599,200$ USD annually.

6.3.4 Statement of the problem

During the year 2002 the scrap was 1.7% in the forming process. Furthermore, 40% of this value was found in products D24 (D stands for rectangle side length, see Figure 6.1) and D30. Therefore, the problem is defined as *a high amount of scrap in the forming process*, and the problem focused is that *40% of the high amount of scrap is found in products D24 and D30* (see Figure 6.2, where a Pareto diagram – a tool to pinpoint the significant elements of the analyzed situation – was used to help focus the main components of the scrap problem).

6.3.5 Project goal

To apply the Six Sigma methodology in the process of product forming to reduce the current overall scrap level from 1.7% (based on year 2002) to a value no greater than 0.63%. This goal is specific, attainable, relevant, and time

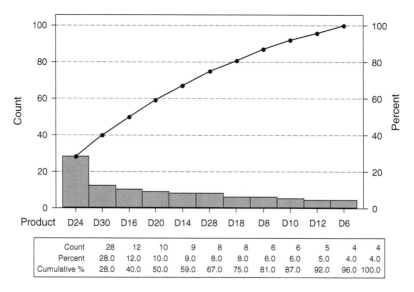

Product	D24	D30	D16	D20	D14	D28	D18	D8	D10	D12	D6
Count	28	12	10	9	8	8	6	6	5	4	4
Percent	28.0	12.0	10.0	9.0	8.0	8.0	6.0	6.0	5.0	4.0	4.0
Cumulative %	28.0	40.0	50.0	59.0	67.0	75.0	81.0	87.0	92.0	96.0	100.0

Figure 6.2 Pareto chart relating scrap vs. product type.

bound (in this case it was limited to 1 year). It is attainable because during a certain month this scrap level was reached. It is relevant because it is tied to the company's operational performance.

6.3.6 Project scope

This project focuses on the process of product forming, because it was here where the greater level of scrap was detected, and is limited to products D24 and D30. The allotted time frame will be no more than a year.

6.3.7 Team formation

The core team members were selected mainly based on their process knowledge, and not on their knowledge of the Six Sigma methodology. They were informally trained as the project developed. They were the quality manager (team leader), the production manager, the quality engineer (green belt), the maintenance engineer, the operator of the press, and the first author of this study as an external coach.

6.3.8 Milestones and communications plan

The project time frame was limited to 1 year. It was developed based on weekly meeting of team members, and occasionally other key personnel in case some specific or additional information or recommendations were needed. The upper management was continuously informed about the progress of this study.

6.3.9 Business process mapping

Figure 6.3 presents a self-explanatory SIPOC diagram for the process of product forming where its suppliers (S), their inputs (I), the selected process (P), its outputs (O), and its customers (C) are described.

Suppliers	Inputs	Process	Outputs	Customers
Suppliers of raw materials Suppliers of the equipment Human resource department Planning department	Raw materials Equipment Personnel Production program	Product forming	Final product Scrap Customers' complaints	Final customers Company quality department

Figure 6.3 SIPOC diagram.

6.4 Measure phase

During this phase the measurement system of the response variable is analyzed, and the process baseline is established.

6.4.1 Measurement system analysis

The next step before moving into the analysis phase is to evaluate the measurement system of the response variable (Y = opposing force in kg), to assess if its measurements are reliable. This was done through a gage repeatability and reproducibility or GR&R study using the preferred analysis of variance or ANOVA method, instead of the X-bar–R (averages and ranges control chart) one, because the last does not allow the evaluation of the interaction term between pieces and operators (Montgomery and Runger, 1993; MSA, 2002). The study was performed by two operators taking 10 different samples of opposing force readings as amperage variation through a transducer, and repeating them three times each. See the appendix for the complete results.

The ANOVA method tests for differences between samples, operators, and the interaction between them and the samples. The analysis of the samples indicates that there is a statistically significant difference between at least one of them and the others at the 5% significance level (the *p*-value, or probability of obtaining a value that extreme in case Ho is true, where Ho is established as no difference between all the samples, is less than 5%). The ANOVA did not detect a statistically significant difference between the operators nor a possible interaction between them and the samples at the 5% significance level (*p*-values of 0.104 and 0.94, respectively).

The statistically significant differences between the 10 samples taken mean that the samples are different and, therefore, appropriate for this study, because it is important to select them to cover all the operating range of the gage (MSA, 2002). This is also reflected in the part-to-part variation below.

In case the difference between the operators had been statistically significant, this would mean that they are not measuring the same way and they should be retrained to perform equally. Similarly, if the interaction between operators and samples would have been statistically significant, this would mean that some operators are having problems measuring some samples and some others not (MSA, 2002). This is not the case. Therefore, the ANOVA table is recomputed to delete the interaction term, and the final analysis does support too the above conclusions. (See the appendix. *Note*: It seems in the ANOVA table that the sums of squares (SS) for operator, samples*operator, and repeatability are zero but this is not the case. They have such small values that they are not shown in the Minitab® analysis. For instance, the current mean squares for repeatability is approximately 0.000814816.)

Next, a % GR&R value of 0.09 (based on the % study variation) was obtained, concluding that the response variable evaluations are reliable. (According to the MSA (2002) and previous editions, if the % GR&R is less than 10, the measurement system is appropriate to produce reliable measurements. If it is between 10% and 30% it depends on the importance of the characteristic being measured. If it is greater than 30 do not use it.)

Similarly, the difference between operators (reproducibility) is 0.01%, and the variation due to the measurement instrument (repeatability) is 0.09%. The number of different categories or equipment discrimination is its ability to detect similar units. This value should be at least 5.

Finally the part-to-part variation has the larger contribution to variation (almost 100%) meaning that the samples are different and, therefore, appropriate for this study; and that the variation between the operators, and the variation due to the interaction term, is negligible.

6.4.2 Process baseline

To obtain the process baseline or the initial estimate of the process capability, a data collection was performed, and the following information was obtained for products D24 and D30 during the second semester of 2002.

During that time period, 20 batches of D24 and 15 batches of D30 were produced. From every batch a variable number of products is obtained. Table 6.1 shows the baseline analysis for product D24. Several batches were omitted in the analysis because their sample sizes were small.

The opposing force is continuously changing, and samples are taken every minute. The first step was to evaluate stability of the opposing force in each analyzed batch by using individuals and moving range (I–MR) control charts (Montgomery, 2001a). In all but two batches the opposing force did not show stability (not shown), although in all cases normality was not rejected by either the Anderson–Darling, Ryan–Joiner or Kolmogorov–Smirnov tests at the 5% significance level (not shown).

Next, the performance indices Pp and Ppk were computed as well as their associated sigma values. Pp and Ppk are commonly used when the process is not stable as this is the case for most of the batches in Table 6.2. They are equivalent to the capability indices Cp and Cpk (that evaluate process variation and centering with respect to its specification limits) except that the previous is computed using the overall sample standard deviation (and therefore represent the current process behavior), and the last is computed using the within (internal) or short-term estimation of process variation. The difference between the 'overall sigma' value and 'Z benchmark' in Table 6.2 is that the first indicates the current process performance, and the last is what is expected to happen if the process variation is reduced to the current minimum attainable level, although it does not mean that this level could not be further reduced if appropriate actions are taken.

Figure 6.4 presents an I–MR control chart for the overall sigma values in Table 6.2 for assessing and confirming their stability.

Then, with the purpose of computing a confidence interval (CI) for the overall sigma values, a normality test was applied to this small sample. Figure 6.5 shows two type of normality test. One is graphical and the other is analytical. Although in Figure 6.5 the data do not follow a perfect straight line, they are

Table 6.1. Relationship between sigma level and ppm

Sigma level	ppm
6	3.4
5	233
4	6210
3	66,807
2	308,537
1	690,000

Table 6.2 Sigma values for the batches sampled

Batch	Samples	Stability	Normality	Pp	Ppk	Overall sigma	Z benchmark
D24-1	50	No	Yes	1.31	1.10	3.29	3.52
D24-2	75	No	Yes	1.01	0.99	2.81	3.58
D24-3	237	No	Yes	0.88	0.64	1.90	2.71
D24-4	172	Yes	Yes	1.13	0.68	2.04	3.23
D24-5	85	Yes	Yes	1.00	0.77	2.31	4.19
D24-6	97	No	Yes	1.10	0.83	2.50	3.82
D24-9	54	No	Yes	1.03	0.45	1.36	2.10
D24-11	151	No	Yes	1.14	1.12	3.22	4.66
D24-13	105	No	Yes	0.55	0.26	0.77	1.64
D24-14	63	No	Yes	0.86	0.65	1.93	2.35
D24-15	119	No	Yes	0.85	0.75	2.18	3.50
D24-16	162	No	Yes	0.86	0.74	2.18	2.78
D24-20	66	No	Yes	1.15	0.83	2.49	3.07

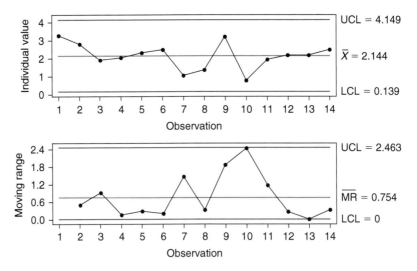

Figure 6.4 Stability evaluation of overall sigma values. LCL: lower specification level; UCL: upper specification level.

not that far either. Nonetheless, if there is any doubt, the Anderson–Darling normality analytic test (see Shapiro, 1990) is shown too. AD stands for Anderson–Darling test statistic (0.248), and its corresponding p-value is 0.696, concluding that the normality of this set of data is not rejected. A p-value or probability of

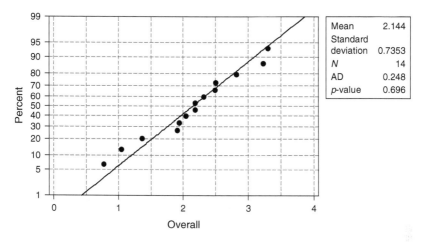

Figure 6.5 Normality test for the overall sigma values.

Table 6.3 Summary of process baseline calculations

Product	95% CI sigma	Mean sigma	Mean ppm	Target	USL	LSL
D24	(1.72, 2.57)	2.144	16,180	32	38	26
D30	(1.54, 2.45)	1.996	22,750	16	19	13

LSL: lower specification level; USL: upper specification level.

obtaining a value that extreme in case Ho (the null hypothesis) is true, where Ho is established as the data set coming from a normal distribution is greater than 5%, and therefore Ho (normality) is not rejected.

Finally, a 95% CI (based on the t-distribution for small samples drawn from a normal distribution) for the mean overall sigma values was computed (1.71973, 2.56884), meaning that with a 95% confidence the 'real' mean overall sigma value (or process baseline for product D24) will lie between 1.72 and 2.57. Applying a similar procedure (not shown), the process baseline 95% CI for product D30 is between 1.54 and 2.45 in sigma units. See Table 6.3 for a summary of the previous calculations. The parts per million (ppm) values were obtained from a standard normal table (see, for instance, Breyfogle III, 1999).

6.4.3 Process description

A detailed description of the forming process – including process and product characteristics – is shown in Figure 6.6. The forming process steps are: the reception and storage of RM where its quality certificate is checked. Next, the RM is sieved and inspected, and the mixing operation is prepared and performed by adjusting process conditions such as feeding speed and temperature. The mixture is then cooled, and the final product if formed (extruded) by a press. Finally, the product is inspected and stored. For the meaning of the symbols in

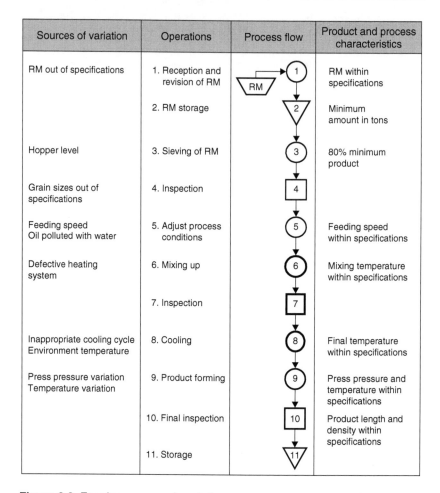

Sources of variation	Operations	Process flow	Product and process characteristics
RM out of specifications	1. Reception and revision of RM	1 / RM	RM within specifications
	2. RM storage	2	Minimum amount in tons
Hopper level	3. Sieving of RM	3	80% minimum product
Grain sizes out of specifications	4. Inspection	4	
Feeding speed Oil polluted with water	5. Adjust process conditions	5	Feeding speed within specifications
Defective heating system	6. Mixing up	6	Mixing temperature within specifications
	7. Inspection	7	
Inappropriate cooling cycle Environment temperature	8. Cooling	8	Final temperature within specifications
Press pressure variation Temperature variation	9. Product forming	9	Press pressure and temperature within specifications
	10. Final inspection	10	Product length and density within specifications
	11. Storage	11	

Figure 6.6 Forming process description.

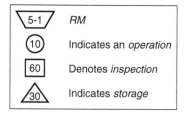

5-1	*RM*
10	Indicates an *operation*
60	Denotes *inspection*
30	Indicates *storage*

Figure 6.7 Symbols explanation.

Figure 6.6, please refer to Figure 6.7, where the number inside them is for illustration purposes only.

The sources of variation in each process step were obtained through brainstorming, and the resulting sources/causes were classified in the Ishikawa diagram depicted in Figure 6.8.

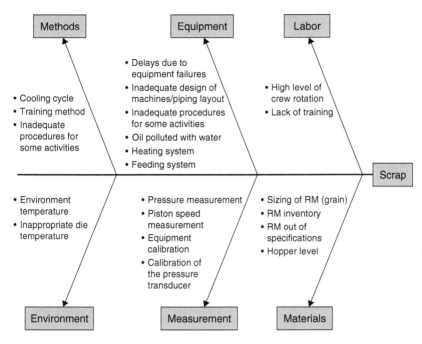

Figure 6.8 Ishikawa diagram.

6.5 Analyze phase

Next, in the analysis phase the potential critical variables are found and analyzed, and the key ones are then confirmed through statistical analyses like test of hypothesis, ANOVA, and others.

First, a failure mode and effects analysis (FMEA) was done to analyze the forming process and its variables. An FMEA is a prevention tool to improve a process by, based on a deep analysis of what may go wrong on every step of the process, implementing preventive mechanisms for potential failures, their causes, or their effects (Stamatis, 1995; FMEA, 2001). The complete analysis is not shown. Part of it is presented in Figure 6.9 for the resulting potential critical factors, and by the improvement actions in the last two DMAIC phases, that is *improve* and *control*.

For clarification purposes, Figure 6.1 is revisited and reproduced as Figure 6.10. The last operation that takes place in the forming process is when the product is extruded by means of a piston moving at a constant speed (operation number 9, pressing, in Figure 6.6). The resistance or opposing force is a function of the product's consistency or density, which in turn is dependent on the number and size of six types of particles (A–F).

Therefore, it is necessary to find the appropriate variables to make the necessary process adjustments to maintain the opposing force as constant as possible all along the production cycle. Then it was decided to perform a 2^k factorial design

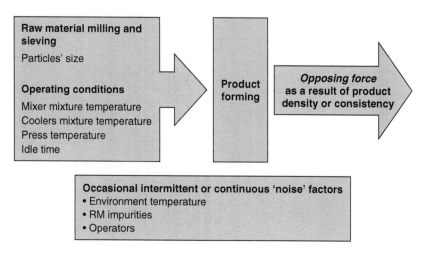

Figure 6.9 Variables in product forming.

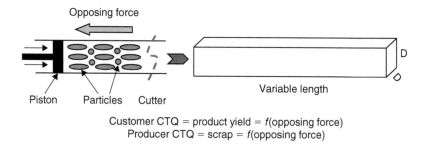

Figure 6.10 The forming process extrusion.

to find the variables that affect press opposing force. A designed experiment is a set of different combinations that are tested to find the statistically significant effects over a response variable, and to increase the knowledge about the process or operation being analyzed (see Antony, 1999).

After several considerations based on the confidential product formulation, three components/variables were chosen to be included in the factorial design. They were the size of particles A, B, and C at two levels each. The goal of this two-stage experimentation was to determine which of the three tested factors, and their interactions, affect the mean (on the first-stage experimentation analysis) and the variation (on the second one) of the response variable, namely *opposing force*.

Therefore, since all the three factors were defined as having two levels (levels are not shown due to confidentiality), the appropriate design is a 2^3 factorial in eight experimental runs. Although the minimum recommended number of replicates – for studying variation effects – is 5 (Schmidt and Launsby,

1994), due to some constraints because of the production schedule, the number of replicates was defined to be two per every experimental condition. Note that due to time constraints, this analysis was applied to product D24 only. Nonetheless, all the different products are differentiated by their dimensions and not by their particle composition.

The initial analysis for the mean opposing force is:

Estimated effects and coefficients for OPPOSING FORCE (coded units)

Term	Effect	Coef	SE Coef	T	P
Constant		32.125	0.2795	114.93	0.000
ParticleA	-3.000	-1.500	0.2795	-5.37	0.001
ParticleB	-0.500	-0.250	0.2795	-0.89	0.397
ParticleC	-0.500	-0.250	0.2795	-0.89	0.397
ParticleA*ParticleB	0.750	0.375	0.2795	1.34	0.217
ParticleA*ParticleC	0.750	0.375	0.2795	1.34	0.217
ParticleB*ParticleC	-2.250	-1.125	0.2795	-4.02	0.004
ParticleA*ParticleB* ParticleC	1.500	0.750	0.2795	2.68	0.028

S = 1.11803   R-Sq = 87.77%   R-Sq(adj) = 77.06%

ANOVA for OPPOSING FORCE (coded units)

Source	DF	Seq SS	Adj SS	Adj MS	F	P
Main Effects	3	38.000	38.000	12.667	10.13	0.004
2-Way Interactions	3	24.750	24.750	8.250	6.60	0.015
3-Way Interactions	1	9.000	9.000	9.000	7.20	0.028
Residual Error	8	10.000	10.000	1.250		
Pure Error	8	10.000	10.000	1.250		
Total	15	81.750				

The final analysis is deleting the non-statistically significant factors, except those who may alter the hierarchy principle where it is established that, if a high-order term is included in a model, it should include the low-order terms too (Montgomery, 2001b):

Estimated effects and coefficients for OPPOSING FORCE (coded units)

Term	Effect	Coef	SE Coef	T	P
Constant		32.125	0.3010	106.71	0.000
ParticleA	-3.000	-1.500	0.3010	-4.98	0.001
ParticleB	-0.500	-0.250	0.3010	-0.83	0.426
ParticleC	-0.500	-0.250	0.3010	-0.83	0.426
ParticleB*ParticleC	-2.250	-1.125	0.3010	-3.74	0.004
ParticleA*ParticleB* ParticleC	1.500	0.750	0.3010	2.49	0.032

S = 1.20416;   R-Sq = 82.26%;   R-Sq(adj) = 73.39%

ANOVA for OPPOSING FORCE (coded units)

Source	DF	Seq SS	Adj SS	Adj MS	F	P
Main Effects	3	38.000	38.000	12.667	8.74	0.004
2-Way Interactions	1	20.250	20.250	20.250	13.97	0.004
3-Way Interactions	1	9.000	9.000	9.000	6.21	0.032
Residual Error	10	14.500	14.500	1.450		
Lack of Fit	2	4.500	4.500	2.250	1.80	0.226
Pure Error	8	10.000	10.000	1.250		
Total	15	81.750				

Based on the individual t-tests (T), the significant effects (P = p-value < 0.05) are particle A, the interaction between particles B and C, and the uncommonly statistically significant triple interaction between all the particles. Additionally, due to the hierarchy principle, the other effects to be included in the model are particles B and C, even though their corresponding p-value is not less than 0.05.

A plausible explanation for the significance of the triple interaction is that the three particles A, B, and C, plus others, constitute the final product, and it is based on their shape and size characteristics that the product consistency is achieved. Furthermore, after obtaining the particles B and C, an extra milling operation is applied to either particle (B or C) to obtain a smaller particle, namely particle A.

The R-Sq, or the coefficient of determination, defines the amount of variation explained by the model; 73.39% is a good figure because it is relatively close to 100%, the maximum attainable value for R-Sq. Note that R-Sq (adjusted) was used instead of R-Sq because the last might be artificially inflated by adding non-significant factors to the model. The R-Sq (adjusted) has a correction factor which takes into account the possibility of overestimating it, and therefore is a more accurate and reliable assessment of the amount of variation explained by the model.

The ANOVA table analyzes the effects as groups of them, that is the p-value = 0.004 for the main effects mean that at least one of the three main effects is significant. Next, the t-tests for the individual effects confirm that, precisely, one of the three main effects is significant, namely particle A. A similar interpretation is valid for the group of three double interactions. The ANOVA analysis for the triple interaction is exactly equal to the individual t-test because there is a single element in this group.

An additional hypothesis test shown in the ANOVA table is the lack-of-fit test, used to assess how well the resulting model fits the data. In this case, a p-value of 0.226 (>0.05) indicates that the null hypothesis – which in words is 'the proposed model fits the data well' – is not rejected.

Finally, it is necessary to perform a last set of tests to check the model's assumptions of normality, constant variance, and independence of the error terms by means of the residual analysis. A residual is the difference between the real 'y' observed value, and its estimation by the model. The normality test is shown in Figure 6.11. As it was explained in Figure 6.5, based on the p-value = 0.709, this test is not significant, meaning the normality hypothesis (Ho) is not rejected.

Figure 6.12 presents the graphical test of the model's assumption of constant variance. For this assumption not to be rejected, the graph should show a

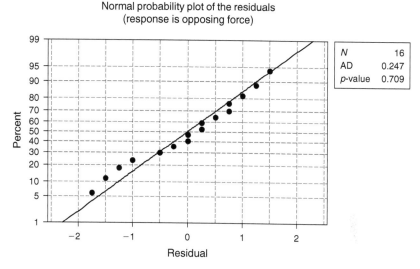

Figure 6.11 Normality test.

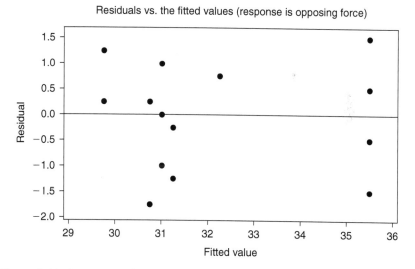

Figure 6.12 Constant variance test.

horizontal band, meaning that the variation between the two replicates for every experimental condition is the same. The test is clearly rejected when the plotted dots have a funnel-shape behavior, which in fact is not the case (see Box and Bisgaard, 1995). Hence, the assumption of constant variance is not rejected either. Unfortunately, due to the way in which the data were obtained, and on process constraints, the order of the data was not clearly defined, and the independence test was not done.

The second stage in the analysis was to assess the influence of the tested effects on the variability of the opposing force. This was done by obtaining the standard deviation of opposing force for every experimental run, and taking it as the new response variable.

The final hierarchical model for the standard deviation (s) of opposing force is:

Estimated effects and coefficients for s (coded units)

Term	Effect	Coef	SE Coef	T	P
Constant		0.8839	0.2165	4.08	0.015
ParticleA	-0.0000	-0.0000	0.2165	-0.00	1.000
ParticleC	-0.0000	-0.0000	0.2165	-0.00	1.000
ParticleA * ParticleC	-1.0607	-0.5303	0.2165	-2.45	0.070

S = 0.612372; R-Sq = 60.00%; R-Sq(adj) = 30.00%

ANOVA for s (coded units)

Source	DF	Seq SS	Adj SS	Adj MS	F	P
Main Effects	2	0.000	0.000	0.0000	*	*
2-Way Interactions	1	2.250	2.250	2.2500	6.00	0.070
Residual Error	4	1.500	1.500	0.3750		
Pure Error	4	1.500	1.500	0.3750		
Total	7	3.750				

The ANOVA table do not show the F-value for the main effects. This is because none of them were statistically significant. Only the interaction term between particles A and C was significant at the 10% significance level. See also the individual t-tests (T). In this case there is a great difference between the R-Sq and R-Sq (adjusted) meaning that the proposed model that includes the significant interaction term between particles A and C, as well as its individual components (particles A and C) due to the hierarchy principle, is over-fitted. According to Montgomery (2001b), it is possible to disregard the hierarchy principle if the resulting model is better.

Therefore, a regression analysis was done to get the non-hierarchical model of s (pA * pC stands for the interaction term between particles A and C):

The regression equation is:

s = 0.884 - 0.530 (pA) * (pC)

Predictor	Coef	SE Coef	T	P
Constant	0.8839	0.1768	5.00	0.002
pApC	-0.5303	0.1768	-3.00	0.024

S = 0.5; R-Sq = 60.0%; R-Sq(adj) = 53.3%

ANOVA

Source	DF	SS	MS	F	P
Regression	1	2.2500	2.2500	9.00	0.024
Residual Error	6	1.5000	0.2500		
Total	7	3.7500			

No evidence of lack of fit (P >= 0.1).

Even though the adjusted R-Sq has increased to 53.3%, a much better value than the previous 30%, it is still not large enough to consider that this model explains a vast part of the variation in Y. It is possible that some other factors that may affect this response variable were not included in the model. Further investigation is needed. Additionally, the assumptions of normality, constant variance, and lack of fit are met (the two first not shown).

6.6 Improve phase

In this phase, the relationship between the independent factors (Xs) and the response variable (Y) is established, and the Xs are optimized. Additionally, specific actions are taken to improve the process.

The final models for mean and for the standard deviation of opposing force are:

$$\bar{y} \text{ (estimated)} = 32.125 - 1.5A - 0.25B - 0.25C - 1.125BC + 0.75ABC$$

and

$$\hat{s} = 0.8839 - 0.5303AC$$

where A, B, and C stand for particles A, B, and C, respectively.

The estimated mean response (based on the model for average opposing force) is shown in Figure 6.13. The coded factors levels where the estimated mean response is closest to the target value of 32 kg are: $A(-1)$, $B(-1)$, and $C(-1)$.

Therefore, the estimated mean opposing force at $A(-1)$, $B(-1)$, and $C(-1)$ is:

$$\bar{y} \text{ (estimated)} = 32.125 - 1.5(-1) - 0.25(-1) - 0.25(-1)$$
$$- 1.125(-1)(-1) + 0.75(-1)(-1)(-1)$$

$$\bar{y} \text{ (estimated)} = 32.25 \text{ kg}$$

And its associated estimated standard deviation for the opposing force is:

$$\hat{s} = 0.8839 - 0.5303(-1)(-1)$$

$$\hat{s} = 0.3536 \text{ kg}$$

It is straightforward to see that the estimated standard deviation is minimum when either A is at level -1 and simultaneously C is at level -1, or when A is at level $+1$ and simultaneously C is at level $+1$. In this case since the optimum levels to adjust mean opposing force are $A(-1)$, $B(-1)$, and $C(-1)$ (see Figure 6.13), the first combination of levels to minimize the standard deviation of opposing force was used, namely $A(-1)$ and $C(-1)$.

Nonetheless, according to the contour plots in Figure 6.14 (a contour plot is a projection of the response hyper-plane of several dimensions into the two-dimensional space), there might be other possible combinations of factor levels that could give a closest-to-target mean opposing force.

For instance, by using the option of 'response optimizer' of Mintab® (see row 'Cur' and '$y = 32$' in Figure 6.15), it indicates that the exact target value

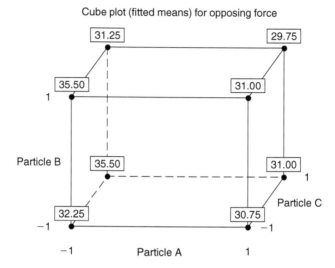

Figure 6.13 Estimation of mean opposing force.

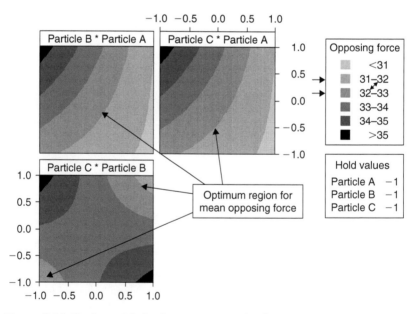

Figure 6.14 Contour plots for the mean opposing force.

is reached at the coded values combination A(0), B(0), and C(0.5) to obtain an estimated response of exactly 32 kg. Nonetheless, the estimated standard deviation would be 0.6189 (almost the double than with the original optimum combination of A(−1), B(−1), and C(−1)).

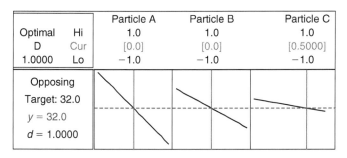

		Particle A	Particle B	Particle C
Optimal	Hi	1.0	1.0	1.0
D	Cur	[0.0]	[0.0]	[0.5000]
1.0000	Lo	−1.0	−1.0	−1.0

Opposing
Target: 32.0
$y = 32.0$
$d = 1.0000$

Figure 6.15 Response optimizer.

Therefore, for the sake of minimizing the variability of the response variable, the selected optimum combination of levels was A (−1), B (−1), and C (−1).

Next, based on the factorial design, and on the FMEA done in the previous phase, the following process improvement actions were carried out to reduce the variation of the particles A, B, and C.

Particle A is obtained from either particle B or C. It also depends on a roller mill that is feed by the RM. The input variables for this operation are the feeding speed, airflow, particle classifier speed, and the size of the RM. It is known that the RM variation of the particles size has a large influence on the final size variation for all particles. To try to reduce this input variation, the appropriated process operating parameters were defined according to the standards of the roller mill equipment manufacture (formerly these parameters were not measured; they were set by operator 'experience' only). Nonetheless, as part of a follow-up project, these recommended levels should be questioned by means of further statistical analyses like design of experiments.

The improvement actions to reduce the size variation of particles B and C were as follows. Since they are the outcome of the operations of milling and sieving (M&S) (see Figure 6.16), they depend on the feeding speed, on the physical conditions of the M&S equipments, and on the size of the RM.

One recommended action to reduce the variation of particles B and C was the addition of a new mill, as well as to replace the hoppers by others of greater capacity to induce less variation.

To summarize the above explanations, Figure 6.17 illustrates the relationship between the analyzed variables; that is the opposing force is a function of the size of the particles A, B, and C, which in turn depend on some other input variables.

6.7 Control phase

In this phase, based on the FMEA, the following necessary improvement and control actions were defined and implemented:

1. An electronic control was introduced for the downloading and storage of RM.
2. Regarding RM sieving, an extra mill was installed and one more has been repaired to reduce cycle time.

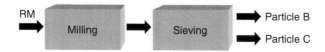

Figure 6.16 M&S operations.

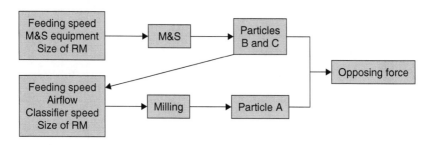

Figure 6.17 Relationship between variables.

3. To reduce variation in the size of particle A, additional controls were installed in a mill to determine the appropriate parameters to make the right selection of grain size.
4. Regulators were installed in each mixer to achieve uniformity in the product.

Finally, two control plans were developed. One is to control the input variables of the milling process where particle A is obtained (see Table 6.4) and the other is for controlling the M&S process from where particles B and C are drawn (see Table 6.5) (the specifications on both the tables are fictitious due to confidentiality). Furthermore, the current forming process' standard operating procedure (SOP) was updated according to the implemented improvement actions described in this section (not shown).

6.8 Economic benefits obtained

To assess the economic benefits of this project, the fraction defective before and after the study was computed. Based on year 2002 the overall initial fraction defective was 1.7%. Now, the fraction defective for product D24 – based on a sample size equivalent to 1635 units of product, from which 28 of them were defective because of the opposing force being outside of its specification limits – was 1.71%. Now, after implementing the recommended improvement actions, approximately 3 months of data were collected for an equivalent sample size of 1021 units of product D24. From these, 10 of them were defective due to the opposing force being outside of its specification limits (see Figure 6.18).

The upper dot plot in Figure 6.18 corresponds to the beginning of this project, and the lower graph corresponds to data obtained after the end of it. Both sets of data were analyzed regarding its stability level and normality status, and did not comply with any of them. Therefore, since the sample sizes are large enough, it is possible to make an empirical estimation of their fraction defective by dividing

Table 6.4 Milling operation control plan for particle A

Parameter	Critical	Specification	Measurement instrument	Responsible	Where	Record/control method	Frequency	Sample size	Reaction plan
Pressure	Yes	3–7 inches of water	Manometer	Operator and quality inspector	Mill	Quality report	Every 2 h	1	Adjust the feeding flow of RM
Classifier speed	Yes	300–350 rpm	Tachometer	Operator and quality inspector	Mill	Quality report and circular graph	Every 2 h	1	Adjust rpm of the classifier according to particle A size
Milling amperage	No	80–90 amp	Ammeter	Operator and quality inspector	Mill	Quality report	Every 2 h	1	Inform maintenance supervisor
Classifier amperage	No	110–120 amp	Ammeter	Operator and quality inspector	Mill	Quality report	Every 2 h	1	Inform maintenance supervisor
Particle A size	Yes	60–70%	Siever	Quality inspector	Laboratory	I–MR chart	Every 2 h	1	Adjust mill's operating conditions

rpm: revolution per minute.

Table 6.5 M&S operations control plan for particles B and C

Parameter	Critical	Specification	Measurement		Where	Record/control method	Frequency	Sample size	Reaction plan
			Measurement instrument	Responsible					
Mill opening	Yes	0–1 inches	Vernier	Operator	Mill	Maintenance report	Whenever required	1	Adjust mill opening according to particles B and C size
Particles B and C size	Yes	95% minimum	Siever	Quality inspector	Laboratory	I–MR chart	Every 2 h	1	Adjust feeding speed or change RM batch or adjust mill opening

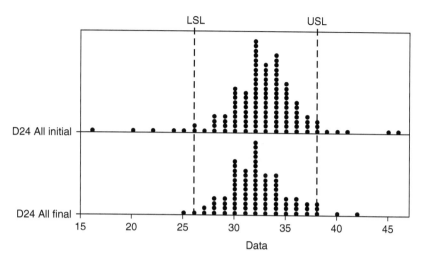

Each symbol represents upto 16 observations.

Figure 6.18 Fraction defective at the beginning and at the end of this study. LCL: lower specification level; UCL: upper specification level.

the number of defective units by the total sample; but, since normality is rejected, the sigma estimates cannot be obtained because they are based on the normal distribution.

Hence the fraction defective after the implementation of the improvement actions is $10/1021 = 0.83\%$. Then, the fraction defective has dropped from 1.71% to 0.83%. Is this fraction defective change statistically significant? To answer this question, a test of hypothesis for the difference of two-proportions was applied:

Sample	X	N	Sample p
1	28	1635	0.017125
2	10	1201	0.008326

```
Difference = p (1) - p (2)
Estimate for difference: 0.00879899
95% CI for difference: (0.000677538, 0.0169204)
Test for difference = 0 (vs not = 0): Z = 2.12; p-Value = 0.034
```

Since the p-value (0.034) is less than the 5% of significance reference value, the null or no difference hypothesis is rejected meaning that the fractions defective before and after the development of this Six Sigma project is statistically significant, the final one being smaller (see the CI where the two limits are positive).

Now, considering all the products, because the improvements affect all of them, the global scrap dropped from 1.7% (in 2002) to 1.2% (in 2003). The corresponding monetary savings are in the order of $280,000 per year (1.7 − 1.2% = 0.5%, then 0.005(952000)/0.017 = 280,000, where $952,000 is the initial COPQ estimated in the business case section). The COPQ dropped from $952,000 to

Table 6.6 Summary of the economic benefits obtained

Product	Initial scrap (%)	Year	Time period	Final overall scrap (%)	Year	Time period	Initial scrap cost	Final scrap cost	p-value
All	1.70	2002	1 year	1.20	2003	1 year	$962,000	$672,000	Less than 5%
D24	1.71	2002	6 months	0.83	2003	3 months	NA	NA	0.034

NA: not applicable.

$672,000 per year. Is this difference in overall scrap level statistically significant? Again a test of hypothesis for the difference of two proportions was applied:

```
Sample              X                    N              Sample p
1                 1024                 60222            0.017004
2                  945                 78734            0.012002

Difference = p (1) - p (2)
Estimate for difference: 0.00500131
95% CI for difference: (0.00371883, 0.00628380)
Test for difference = 0 (vs not = 0): Z = 7.82; p-value = 0.000
```

Since the p-value (0.000, does not mean it is actually zero but a very small value) is less than the 5% of significance reference value the null or no difference hypothesis is rejected, meaning that the overall fraction defective before and after the development of this Six Sigma project is statistically significant, the final one being smaller (see the CI where the two limits are positive). Note that the way in which the overall percentage scrap is obtained is different from the scrap obtained for product D24. The first is the quotient of the number of defective pounds (due to opposing force outside its specification limits) divided by the number of input pounds. The values in the hypothesis test are modified but proportional to the real ones because of confidentiality. See Table 6.6 for a summary of the above explanations.

6.9 Conclusions, comments, and lessons learned

The Six Sigma DMAIC methodology was successfully applied as an isolated case to a process where the variation of RM, in the manufacture of coal products, is significant and its reduction is crucial in achieving final product quality.

The relationship between the response variable (opposing force) and the critical variables was established through two empirical models: one for the mean response and the other for the standard deviation of the opposing force. Then these models were optimized, and the critical factors were controlled. Although the project goal to reduce scrap from 1.7% to 0.63% was not completely met, there is a tendency to achieving it. The overall scrap dropped to 1.2%, and this improvement level satisfied the upper management. On December of this year (2004), an updated information on the 2004 YTD showed an overall scrap level of 0.9%, and the difference between this figure and the year 2003 of 1.2% is statistically significant (test not shown) meaning that the overall scrap level is still dropping, and getting closer to the original goal of 0.63%. The authors of this study still think that the proposed goal is realistic although not in a year's time as previously thought.

Furthermore, independently of the economic benefits obtained, there was some important knowledge gained all along the development of this project. First, several barriers had to be overcome for this methodology to be applied in this company. For instance, certain degree of resistance was present because this initiative did not come from the company's general manager but from the quality manager as a pilot project. Second, there was lack of knowledge about this methodology and a belief that it was not applicable to this kind of processes, or that it was not possible to achieve 3.4 ppm here either.

Additionally, at the beginning of the project it was difficult to gather all the team members due to other assignments they had, until, by means of a company initiative unrelated to this Six Sigma project, it was decided to establish work teams to solve quality and productivity problems. This company initiative was promoted because of a significant production increase.

Also, at the beginning of this project, some team members did not find it necessary to use several analysis tools because they were eager to jump to the solution of the problem. Unfortunately this attitude is frequently found in industry until they are trained and finally understand the required steps a project must undergo for its appropriate development and success. A sensitization process showing the importance of following all the necessary steps of the project overcame this barrier.

The steps followed in the development of this project were:

1. The elaboration of a proposal of where the Six Sigma methodology could be implemented knowing its potential benefits.
2. Before forming the team, and although not as a formal training, the quality engineers support group were presented with this methodology by one of them with certain knowledge of it. From them the team members were selected as well as others from different areas.
3. An agreement between the team members regarding the proposed goal to achieve by using the Six Sigma methodology. The goal was redefined a few times.
4. To have a clear understanding of the process where this project was developed, as well as the team leader's explanation of the advanced statistics tools to be used.
5. The evaluation and approval of the response variable measurement system.
6. Defining of the possible causes of variation of the response variable.
7. Evaluating the effect of the investigated factors, and, based on the critical ones, to develop empirical models for the mean and standard deviation of the response variable.
8. Determine the optimum levels for the critical factors.

9. Define improvement actions (assigning responsibilities and due dates) on the operations from where the critical factors are obtained.
10. Monitor the effects of the implemented improvement actions.

Finally, after the success in the implementation of this methodology, and although it was not officially considered by the upper administration, the team members expected that this project would be the drive for this company to start a larger adoption of Six Sigma. This was not so. Due to budget constraints, this company's upper management decides to put on hold to this initiative. A personal opinion of the first author of this study is that possibly they are not completely convinced on investing on Six Sigma yet, and also maybe because this case study was not suggested by them.

References

Antony, J. (1999). Ten useful practical tips for making your industrial experiments successful. *The TQM Magazine*, 11(4), 252–256.

Bañuelas, R. and Antony, J. (2004). Six Sigma or design for Six Sigma. The *TQM Magazine*, 16(4), 250–263.

Box, G.E.P. and Bisgaard, S. (1995). *Design of Experiments for Discovery, Improvement and Robustness: Going Beyond the Basic Principles* (Notes from the Seminar, March 25–29, 1996). Center for Quality and Productivity Improvement, University of Wisconsin-Madison, USA, pp. 1.4.38–1.4.49.

Breyfogle, III F. (1999). Implementing Six Sigma. *Smarter Solutions Using Statistical Methods*. New York, USA: Wiley-Interscience, pp. 694–695.

Escalante, E. (2003). *Six Sigma. Methodology and Techniques. (Seis Sigma. Metodología y Técnicas). Limusa (in Spanish)*. México, p. 9.

FMEA (2001). DaimlerChrysler Corporation, Ford Motor Company, General Motors Corporation. *Potential Failure Mode and Effects Analysis* (3rd Edition). USA: Automotive Industry Action Group (AIAG), pp. 1–4.

Folaron, J. and Morgan, J.P. (2003). The evolution of Six Sigma. *ASQ Six Sigma Forum Magazine (Milwaukee)*, August, 2(4), 38–44.

Montgomery, D. (2001a). *Introduction to Statistical Quality Control* (4th Edition). New York: Wiley, pp. 249–253.

Montgomery, D. (2001b). *Design and Analysis of Experiments* (5th Edition). New York: Wiley, p. 203.

Montgomery, D. and Runger, G. (1993). Gage capability and designed experiments (Parts I and II). *Quality Engineering*, 6(1), 115–135; 6(2), 289–305.

MSA (2002). DaimlerChrysler Corporation, Ford Motor Company, and General Motors Corporation. *Measurement Systems Analysis* (3rd Edition). USA: Automotive Industry Action Group (AIAG), pp. 43–56, 73–77, 97–98, 117–124.

Schmidt, S. and Launsby, R. (1994). *Understanding Industrial Designed Experiments* (4th Edition). USA, Colorado Springs, CO: Air Academy Press, p. M-2, Appendix M.

Shand, D. (2001). Six Sigma. *Computerworld*, 35(10), 38–41.

Shapiro, S. (1990). *The ASQC Basic References in Quality Control: Statistical Techniques*. The American Society for Quality Control, WI, USA: Milwaukee, pp. 50–52.

Snee, R. (2003). The Six Sigma sweep. *Quality Progress, Frontiers of Quality*, September; 77 (Table 1).

Stamatis, D.H. (1995). *Failure Mode and Effect Analysis. FMEA from Theory to Execution*. USA, Milawaukee, WI: ASQC Quality Press, pp. 25–46.

Trivedi, Y. (2002). Applying Six Sigma. *Chemical Engineering Progress*, July, 76–81.

Waddick, P. (2001). Six Sigma DMAIC Quick Reference. Available at www.isixsigma.com/library/content (accessed 17 November 2004).

Williams, M., Bertels, T. and Dershin, H. (2002). *Rath & Strong's Six Sigma Pocket Guide*. Lexington, MA: Rath & Strong, pp. 5–7.

Appendix

GR&R study – ANOVA method

Two-way ANOVA table with interaction

Source	DF	SS	MS	F	P
Samples	9	126011	14001.2	17183271	0.000
Operator	1	0	0.0	3	0.104
Samples * Operator	9	0	0.0	0	0.940
Repeatability	40	0	0.0		
Total	59	126011			

Two-way ANOVA table without interaction

Source	DF	SS	MS	F	P
Samples	9	126011	14001.2	7298489	0.000
Operator	1	0	0.0	1	0.244
Repeatability	49	0	0.0		
Total	59	126011			

GR&R

Source	VarComp	%Contribution (of VarComp)
Total Gage R&R	0.00	0.00
Repeatability	0.00	0.00
Reproducibility	0.00	0.00
Operator	0.00	0.00
Part-To-Part	2333.53	100.00
Total Variation	2333.53	100.00

Source	StdDev (SD)	StudyVar (6 * SD)	%Study Var (%SV)
Total Gage R&R	0.0441	0.264	0.09
Repeatability	0.0438	0.263	0.09
Reproducibility	0.0050	0.030	0.01
Operator	0.0050	0.030	0.01
Part-To-Part	48.3066	289.840	100.00
Total Variation	48.3066	289.840	100.00

Number of Distinct Categories = 1545

Note: The '0' values are really very small values compared with the large values like 126011 and others. Their real values do not appear in this printout, but they are not zero.

Part II

Applications of Six Sigma in Service Sector

7

Six Sigma in healthcare: a case study with Commonwealth Health Corporation

Lisa Lopez

7.1 Introduction to Commonwealth Health Corporation

Commonwealth Health Corporation (CHC) is a private not-for-profit holding company based in Bowling Green, Kentucky and is the parent company of The Medical Center at Bowling Green. The Medical Center at Bowling Green is a 302-bed facility, whose staff has been serving patients in Southcentral Kentucky for over 77 years. CHC also owns and operates The Medical Center at Scottsville, The Medical Center at Franklin, and Commonwealth Regional Specialty Hospital. The combined facilities offer 542 acute and long-term care licensed beds serving as a regional referral center. CHC offers a full range of acute and tertiary services and generates approximately $350million in revenue each year.

7.2 Why Six Sigma in CHC?

In November 1997, CHC's President and CEO, John Desmarais, attended a conference where Jack Welch, former CEO of General Electric (GE), spoke of his vision for GE as a World-Class Six Sigma Company. It was Mr. Welch's passion for Six Sigma and the robust methodology that initiated the drive for CHC's pursuit of perfection. Mr. Desmarais returned from the conference with certainty that Six Sigma would change the approach to delivering care and would offer a competitive advantage setting CHC apart from any other provider. In the Spring of 1998, CHC announced its partnership with GE and launched an organization-wide initiative to pursue Six Sigma quality.

This passion for quality and pursuit of perfection placed CHC as a front-runner to achieving total organizational transformation. With GE's assistance, CHC began a step-by-step implementation of the process throughout the organization. This implementation was not initiated due to adversity, but rather CHC had experienced success in comparison with its competitors. However, success often makes it more difficult to achieve higher levels of performance due to complacency. Achieving excellence requires a vision to move beyond the present and into the future. Senior leaders committed to increase and maintain employees' motivation to excel in the delivery of care, as well as transforming the corporate culture into one of complete excellence, with the goal to virtually eliminate all process errors.

CHC's leaders have always been committed to quality and toward that end several quality programs have been implemented throughout the years. Each of them has delivered a measure of results, but none entirely fulfilled the organization's objectives for excellence. Six Sigma provides the essential tools and the methodology to augment the existing quality initiatives already in place. These initiatives include quality resource management, customer satisfaction, product line task forces, and outcomes Management.

With the number of different quality measures that must be recorded and reported, it is not difficult to allow the organization's improvement approach to become more reactive rather than proactive. Six Sigma provides the means to ensure patients receive care that is proven effective and is the accepted 'standard.' It also designs processes to enforce the prevention of medical errors or 'unthinkable accidents.' In accordance with CHC's mission and vision, to provide high-quality services with unyielding integrity in a caring, safe, service-oriented and cost-efficient manner, Six Sigma was selected and continues to be the capstone to all quality initiatives.

Many healthcare organizations have begun to view Six Sigma as the next step in improvement methodology. For example, Six Sigma may focus on customer satisfaction through reduced wait time, the quality of care through the reduction of medical errors, timeliness of service by providing results more quickly, and the delivery of care at a lower cost through increased productivity. At CHC, Six Sigma is utilized to focus on three corporate metrics as strategic objectives: customer satisfaction measured by Press Ganey Scores; quality of care and service measured by the Centers for Medicare and Medicaid Services clinical projects and timeliness of service within 16 of the organization's core processes; and cost efficiency measured by the total dollar value of implemented projects. Together, these focal areas are the organizational dashboard used to monitor performance and determine the level of care provided. These are the areas in which CHC's executive vice presidents serve as 'Corporate Sponsors' to drive improvement efforts.

7.3 Six Sigma structure at CHC

Six Sigma holds a prominent place in the organization's structure and is led by CHC's president and CEO. It is his vision that drives the initiative. Reporting to him are three executive vice presidents who each direct the efforts of the corporate metrics they support. Vice presidents provide project sponsorship to project leaders and ensure they are informed of any events that may impact their project.

Brown Belts, who are selected by senior leaders, devote approximately 20% of their time actively leading Six Sigma projects. Trained Green Belts function as a layer of support, providing assistance to Brown Belts who lead projects. Green Belts are given specific tasks throughout the projects in which their help is needed to achieve necessary improvements. In this structure, Green Belts may pursue the position of Brown Belts by submitting a letter of recommendation from their respective vice president, with approval from their executive vice president.

Finally, there are three Master Black Belts who serve as teachers and mentors to the organization, and lead more difficult and complex projects. Additionally, a senior vice president manages project selection, removes barriers to change, communicates project successes, and oversees the entire initiative. Together with change agents who are trained facilitators, the Green Belts, Brown Belts, Master Black Belts, and senior vice president as champion, make up the Six Sigma team. It is the Six Sigma team's charge to maintain the momentum of results that Six Sigma has provided and to make use of resources even more efficiently.

Senior leaders made a commitment to learning at CHC. All employees receive a primer to Six Sigma during new employee orientation. Additionally, every employee receives at least 1 full day of training devoted to Six Sigma, while many may receive more. One critical factor is that all senior managers, including the CEO, have been Green Belt trained and shadowed a project. Belt training is quite rigorous and consists of 13 intense days extending over a 6-month period. Training materials for classes were all developed by CHC's Master Black Belts and have been customized to healthcare. They continually update the material and include many project examples from Belt projects at CHC.

7.4 Six Sigma methodology

Each phase of the DMAIC (Define–Measure–Analyze Improve–Control) methodology includes an assortment of tools that provide further understanding of a process. One of the many strengths of Six Sigma is that every project is approached as the same, utilizing each of the five phases; there are no short-cuts to process improvement in the world of Six Sigma. All analysis is based on actual data, rather than opinion or perception.

At CHC, Six Sigma refers to more than the pure statistical methodology. It is a strategic initiative that includes transforming culture. In addition to teaching data collection and analysis tools, change acceleration process (CAP) helps employees to create a shared need, shape a vision, mobilize commitment, make changes last, and then monitor progress. Both CAP and Six Sigma projects are supported by WorkOut™ (town meetings), which is an operational tool to help cut through bureaucracy in decision-making. It also brings together those most closely involved in a process to drive improvements. Both CAP and WorkOut™ training was provided by General Electric Medical Systems (a division of GE) and equipped the organization to face the challenge of change with a proven method.

7.5 Benefits of Six Sigma

The benefits of Six Sigma projects are experienced throughout the organization. During the initial implementation, the first 'Belt' class focused on processes in radiology. As a result of these projects, costs per procedure were significantly reduced. Additionally, examination results are distributed to ordering physicians faster, patients receive treatments quicker, and the physical workspace was re-designed to increase employees' efficiency.

Other classes have focused on different areas of the organization: maternal care, specific pulmonary diagnosis related groups (DRGs), admissions, the billing process, and the documentation process. Using the skill set learned in Six Sigma, CAP and WorkOut™, and other organizational resources allows managers to tackle issues such as patient satisfaction in the emergency department and medical/surgical areas, patient throughput, employment processes, surgery scheduling, and revenue cycle. Due to the rigorous statistical training of Six Sigma, managers ask for data before making decisions and utilize the tools to assist in performance improvement in their areas.

Six Sigma is a data-driven methodology that is well understood by medical practitioners, making it easier to create buy-in. Physicians utilize the concept of DMAIC in their practices daily, defining a patient's problem, obtaining a baseline measure, analyzing the factors that could cause the illness, prescribing a remedy to improve the patient, and then maintaining control of the patient's health. This is the same method used in studying process problems using Six Sigma methodology. Physicians and other practitioners are more apt to accept change in processes when the decision is based on data. Many times, projects may require a physician champion to implement appropriate solutions successfully. This helps to foster an environment of playing on the same team, which benefits the organization, physicians, and most of all the patients.

Essentially, Six Sigma targets the root cause responsible for process problems and provides the necessary tools to resolve them. The difference in this approach and others is its built-in skill set and monitoring capabilities, proving to be the most effective system-wide quality improvement method focusing on results. Each project represents a significant opportunity to improve some aspect of the services provided by CHC, and Belts realize the extent to which they can positively impact the way CHC does business.

7.6 Critical success factors of Six Sigma in healthcare

Most healthcare organizations would not dispute that there is a need to improve, and the effectiveness of Six Sigma has been proven in the manufacturing environment at well-known companies such as Motorola and GE. CHC's leaders have considered its applicability to healthcare from the perspective of quality of processes, or the ability to deliver care prescribed by practitioners. Together with the quality of practice, or practitioners' judgment, these components drive the effective and efficient delivery of service. Senior leaders at CHC determined that Six Sigma was not the approach to determine the appropriate method of treatment; rather it is ideally suited to designing processes that deliver care in a timely and efficient manner.

Commitment is critical. Applying Six Sigma in a service industry is not easy, and if senior leaders are not on board, it is almost certainly a formula for failure. Selecting the right people on the Six Sigma team is crucial as is the selection of projects, which must be tied to the organization's strategic imperatives. Project ideas are obtained from a variety of sources including feedback from our customer service tools, benchmarking data and analysis, financial results, WorkOut™, and brainstorming sessions. All Six Sigma projects must support one of the metrics and the 'project R0' or charter must be approved by the executive vice president sponsoring that metric. The appropriate project leader, 'Belt,' is assigned based on the scope of the project. One of the Master Black Belts mentors the Belt to ensure whether the methodology is strictly adhered.

In addition to selecting the right project, project success is directly related to buy-in and ownership of processes. The selected project must be important not only to senior leadership but also to the project leaders. If projects have been tied to the areas of strategic focus, there is buy-in from senior leadership. They will in turn ensure the department managers are held accountable to produce the established targets. The management team and the whole organization, not just the project leaders, must own the targets.

Financial results and their validation continue to be a challenge in healthcare. In a manufacturing environment, it is straightforward to measure the cost of production. However, it is very difficult to place a dollar value on a faster test result that may yield a shorter length of stay or the value of a more satisfied customer. In healthcare, there are many projects that are truly worthwhile and impact the quality of care that are not financially driven. Therefore, the organizational focus of projects must be determined early on to ensure the method of project selection is well planned and considerations made for all potential aspects.

It is important to offer an incentive for project leaders as they take on more organizational responsibilities. CHC offers a financial incentive to help motivate Brown Belts. Each project leader is eligible to receive as much as $2500 for a successful project. Payment is tied to three criteria:

1. Timely completion (completion times are agreed upon by the sponsor, Master Black Belt, and project leader at the project's onset).
2. The project must pass each review stage or tollgate (Master Black Belts must sign off on the project methodology).
3. Finally, the project must result in a 70% reduction in the number of defective parts per million (DPPM). At the project's completion, $2000 is paid and then another $500 if the project reaches Six Sigma and improvements have sustained at the next 6-month measure.

CHC has embraced a culture change throughout the organization that has impacted every employee. Resistance is a natural part of change and with change of this magnitude it is imperative to offer a means to overcome resistance and create buy-in from all employees. Barriers to effective communication have been eliminated, the 'silo' mentality between departments has been broken, and employees tackle problems with a more data-driven approach.

7.7 Clinical project example using DMAIC

One critical element of any successful project is the ability to effectively execute it. Tricia Just is the Infection Control Team Leader at The Medical Center in Bowling Green. She has been with CHC for over 28 years. She completed a DMAIC project that focused on surgical site infections. The Infection Control Department monitors selected surgical cases each year. These cases are chosen due to high volume, high risk, and/or are problem prone. Ms. Just found that increased instances of wound integrity problems up to and including deep intra-abdominal infections were being identified in small and large bowel surgical cases. As the Infection Control Group had been unable to identify any particular area for improvement, the Six Sigma project was undertaken to determine the cause and ultimately reduce the infection rate.

7.7.1 Six Sigma project: surgical site infections in bowel cases

Define

As with all projects at CHC, a project charter or R0 (Figure 7.1) was completed, which describes the project and how it will be measured, what will be considered a defect, and how data will be attained. It also requires the signatures of the Six Sigma Master Black Belts as well as the project sponsor to ensure the necessary elements are included and buy-in has been obtained.

The project was identified due to a higher than expected number of post-operative wound infections and related wound problems noted in bowel cases, which led to increased patient days, costs, and morbidity. The response variable or 'Y' measurement included small and large bowel cases (excluding any that had an implantable device, emergency cases, and Class 4 or infected cases). These cases were followed for post-operative wound healing problems up to and including infection, measuring the number of days before a surgical site infection (superficial, deep, or organ space), or wound integrity problem (patient started antibiotics, wound cultured, any drainage purulent or not, and/or wound open and packed) developed within 30 days of surgery.

The project was based on the Centers for Disease Control (CDC) and Prevention's definition of a nosocomial surgical site infection and epidemiological evidence of problems with wound integrity. A nosocomial infection is an infection that was not present or incubating at the time of admission to the hospital. The infection may also be hospital acquired but not become evident until after discharge from the hospital.

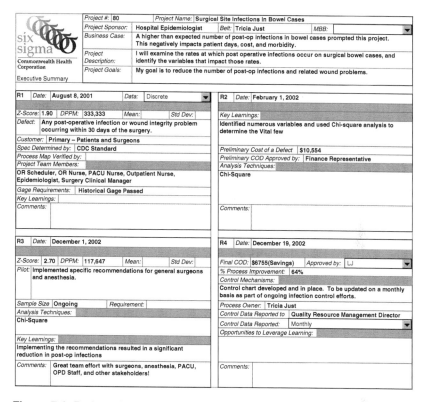

	Project #:	80		Project Name:	Surgical Site Infections in Bowel Cases

six sigma
Commonwealth Health Corporation
Executive Summary

Project Sponsor: Hospital Epidemiologist | Belt: Tricia Just | MBB:

Business Case: A higher than expected number of post-op infections in bowel cases prompted this project. This negatively impacts patient days, cost, and morbidity.

Project Description: I will examine the rates at which post operative infections occur on surgical bowel cases, and identify the variables that impact those rates.

Project Goals: My goal is to reduce the number of post-op infections and related wound problems.

R1 Date: August 8, 2001   Data: Discrete

Z-Score: 1.90 DPPM: 333,333 Mean: Std Dev:

Defect: Any post-operative infection or wound integrity problem occurring within 30 days of the surgery.

Customer: Primary – Patients and Surgeons

Spec Determined by: CDC Standard

Process Map Verified by:

Project Team Members:
OR Scheduler, OR Nurse, PACU Nurse, Outpatient Nurse, Epidemiologist, Surgery Clinical Manager

Gage Requirements: Historical Gage Passed

Key Learnings:

Comments:

R2 Date: February 1, 2002

Key Learnings:
Identified numerous variables and used Chi-square analysis to determine the Vital few

Preliminary Cost of a Defect $10,554

Preliminary COD Approved by: Finance Representative

Analysis Techniques:
Chi-Square

Comments:

R3 Date: December 1, 2002

Z-Score: 2.70 DPPM: 117,647 Mean: Std Dev:

Pilot: Implemented specific recommendations for general surgeons and anesthesia.

Sample Size Ongoing   Requirement:

Analysis Techniques:
Chi-Square

Key Learnings:
Implementing the recommendations resulted in a significant reduction in post-op infections

Comments: Great team effort with surgeons, anesthesia, PACU, OPD Staff, and other stakeholders!

R4 Date: December 19, 2002

Final COD: $6755(Savings)   Approved by: LJ

% Process Improvement: 64%

Control Mechanisms:
Control chart developed and in place. To be updated on a monthly basis as part of ongoing infection control efforts.

Process Owner: Tricia Just

Control Data Reported to Quality Resource Management Director

Control Data Reported: Monthly

Opportunities to Leverage Learning:

Comments:

Figure 7.1 Project charter.

Measure

The project measurement was discrete, and the number of defects and opportunities were determined. A team was formed of nurses from the operating room (OR), out-patient department, surgical unit, and post-anesthesia care unit (PACU) along with a scheduler. The hospital's epidemiologist and OR clinical manager served as consultants and champions for the project. Process maps (Figures 7.2 and 7.3) were developed for the pre-operative period and for the surgical day through to discharge utilizing the expertise of the team. Staff members, to ensure accuracy in each step, validated these maps.

The Infection Control Team Leader obtained medical records of bowel cases for the first 6 months of the calendar year (Figure 7.4). These records were reviewed to determine variables affecting the process as well as any documentation on infections. Surgeon's offices were contacted for protocols on pre-operative antibiotics and bowel preparation to compare with the current practice.

A historical Gage R&R (repeatability and reproducibility) was performed with 100% accuracy, ensuring the data was collected properly (Figure 7.5). The Brown Belt collected the data and then tested two additional staff members

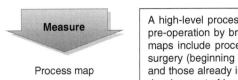

Measure

Process map

A high-level process map was developed for pre-operation by brainstorming with the team. The maps include processes for patients coming in for surgery (beginning with the physician scheduling it) and those already in the hospital (beginning with the development of bowel problems requiring a workup).

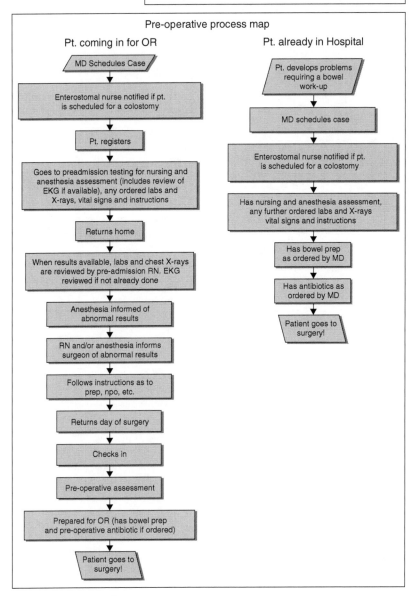

Figure 7.2 Pre-operative process map.

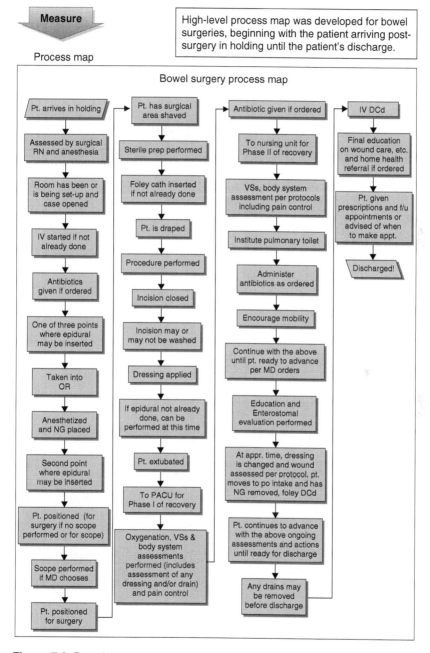

Figure 7.3 Bowel surgery process map.

using three medical records to make certain that they extracted the same information. The process was measured and determined a mean of 5.65 days and standard deviation of 2.4 days. Only the defects were included in calculation of the mean and standard deviation. The upper specification limit was 30 days.

Data collection

- Pulled medical records for the first 6 months of calendar year 2000.
- Reviewed the medical records for variables and any documentation on infections.
- Surgery Team member has reviewed documentation on 'Flash' records, type of drape used, and whether powdered gloves were worn by the assistant and surgeon.
- Surgeon's offices were contacted for protocols on pre-operative antibiotics and bowel preparation.

There are numerous trivial many X's in the process.

Figure 7.4 Data collection.

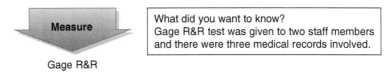

Gage R&R

- I collected the data myself but tested others on their ability to score the medical record the same as me.
- Developed data collection table with several variables.
- Utilized two additional staff members to review three medical records.
- I tested whether they could extract the same information from the medical record as me.
- They scored 100% on the Gage!

Figure 7.5 Gage R&R.

A normality test was run to determine if the sample/population was normally distributed (Figure 7.6). The data was not normal, with a p-value <0.05. It also would not normalize using Box Cox transformation. Therefore, a product performance report was used to calculate a Z-score. This indicated a Z-score of 1.931 and DPPM of 333,333. Simply put, based on the sample, out of 1 million bowel cases, 333,333 would be considered defects (Figure 7.7).

Analyze

The team also brainstormed variables affecting the process and identified the 'trivial many' with the goal of narrowing to the vital few. These variables included: discharge instructions, day of week, time of day, flashed instruments used, prophylactic antibiotics, bowel preparation, surgeon and others on case, shave preparation, surgical preparation, patient risk factors, reason for surgery, type of drape used, skin closure used, drains used, irrigant and how much,

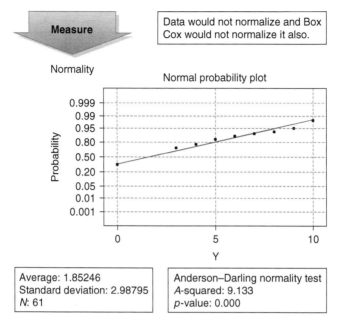

Figure 7.6 Normality test.

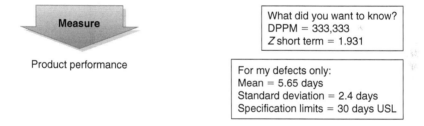

Report 7: Product performance

Characteristic	Defects	Units	Opportunities	Total opportunities	DPU	DPO	PPM	Z-shift	Z-bench
1	20	60	1	60	0.333	0.333333	333333	1.500	1.931
Total	20			60	0.333333	333333	1.500	1.931

Figure 7.7 Product report.

dressing used, post-operative unit, complications, when out of bed, discharge status, powdered gloves used, oxygen use, home with staples/suture in, type procedure, number of procedures, Hibiclens bath pre-operatively, length of stay, length of case, and many others.

Chi-Square test

Chi-Square test: CombFail, CombPass

Expected counts are printed below observed counts

	CombFail	CombPass	Total
1	20	40	60
Before	13.73	46.27	
2	7	51	58
After	13.27	44.73	
Total	27	91	118

```
Chi-Sq = 2.865 + 0.850 +
         2.963 + 0.879 = 7.557

DF = 1, P-Value = 0.006
```

What did you learn?
There is something different about the X's as indicated by the *p*-value <0.05. There is a statistically different (lower) rate of post-operative nosocomial infections after the improvements were implemented.

Figure 7.8 Chi-Square test.

Chi-Square test was used to help determine which of the variables impacted the process the most. This analysis compares discrete data of one variable at two or more levels. Those with a *p*-value of <0.05 are statistically significant. The project analysis identified the following 'vital few' from the trivial many variables:

- Number of procedures performed.
- Type of dressing applied post-operatively (dry dressing vs. betadine vs. bacitracin/bactroban).
- Betadine dressing vs. bacitracin/bactroban.
- Oxygen use on post-operative unit (none, liter flow, concentration).
- Oxygen use on post-operative unit (<40% vs. >40%).
- Hibiclens bath pre-operatively.
- Discharge to self-care vs. Home Health vs. other facility*.
- Discharge to self-care vs. other*.

Two of the vital few were items that could not be addressed by the Brown Belt. One was the number of procedures performed during the surgery. If a patient needs multiple surgeries at one time, all procedures must be performed rather than returning the patient to surgery. It is well documented that multiple surgeries lead to a higher infection rate. This is another variable that showed significance for those who were discharged to other facilities or home care. However, all but two of the defect cases had problems developing while in the hospital and required assistance after discharge (Figure 7.8).

Following the elimination of the other two variables, three significant variables were left – use of Betadine on the post-operative wound, oxygen concentration post-operatively, and Hibiclens bath pre-operatively.

---

*Note that all but two patients had problems while still in hospital.

Cost analysis   The vice president of finance reviewed all the accounts within the Brown Belts dataset and researched cost, charges, reimbursement and insurance group for this set of patients. A distinction was made between those patients who developed an infection and those who did not, and then average cost, average charges, average reimbursement, and average profit for each set of patients was calculated. Based on these calculations, the financial impact per defect was $263.87 (20 defects for 6 months). The annualized cost of a defect was $10,554. All of the infections identified during the study period were superficial infections.

Other considerations were that deep and organ space infections occur and would have a higher financial impact since they require more intensive treatment and care. Additionally, more private payers are moving away from per *diem driven* reimbursement toward DRG reimbursement. Last but certainly not the least, the patient satisfaction factor is immeasurable as the cost associated with the satisfied customers is difficult to quantify.

Improve

Based on the statistical analysis, a pilot was recommended to confront the most significant vital few identified. The recommendations included:

- Hibiclens shower within 48-h pre-operatively,
- dry sterile dressing or only Bactroban or Bacitracin ointment,
- >40% oxygen in the OR,
- >40% oxygen during the immediate 2-h post-operative period.

These specific recommendations were presented to general surgeons and anesthesiology via a letter from the hospital's epidemiologist and the Brown Belt. Additionally, the Brown Belt met several times with the chief of anesthesiology regarding oxygen utilization in the OR and in PACU. The pilot was implemented and then the process was re-measured to determine if a statistically significant improvement was made. Chi-Square test showed that there was a statistically lower rate of post-operative nosocomial infections after the improvements were implemented.

Due to the relatively small number of cases performed, several months of data were obtained to prove a statistical significance based on sample size (Figure 7.9). The product report showed a significant improvement from the baseline measurement of 1.931 Z-score to a Z-score of 2.672. This indicates that out of 1 million opportunities, 120,690 would be considered defects, compared to 333,333 in the baseline dataset. The pilot yielded a 12% infection rate, reducing defects by 21% (Figure 7.10).

Control

A control chart was established to monitor the process and determine if the improvements were maintained (Figure 7.11). A final review of the annualized cost of a defect was performed by finance. The annualized cost was $10,554. The project resulted in a 64% improvement, which saved $6755 annually

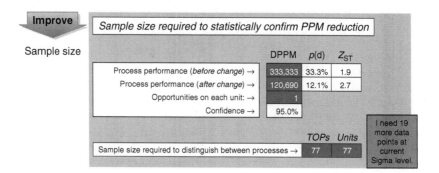

- There is a statistically significant reduction in the number of post-operative infections.
- Due to the relatively small number of cases performed, it will be 2–3 more months before I can claim statistical significance regarding sample size.
- However, because bowel cases are part of ongoing infection control surveillance, I will be aware of any significant deviations over time.
- A control chart has been established to monitor this process and determine if the improvements are maintained.
- Immediate action will be taken if the process shows signs of going out of control.
- Utilizing a closely monitored control program will allow me to finish this project and begin another focus.

Figure 7.9 Sample size.

Product report

Report 7: Product performance

Characteristic	Defects	Units	Opportunities	Total opportunities	DPU	DPO	PPM	Z shift	Z bench
Historical	20	60	1	60	0.333	0.333333	333333	1.500	1.931
New (May–November 2002)	7	58	1	58	0.121	0.120690	120690	1.500	2.672
Total	27			118		0.228814	228814	1.500	2.243

Figure 7.10 Product report.

(Figures 7.12 and 7.13). The Infection Control Team monitors the process and takes action as indicated.

After implementing controls during the ongoing monitoring, the infection rate was 13% and has stayed at that level. For national comparison, the goal is

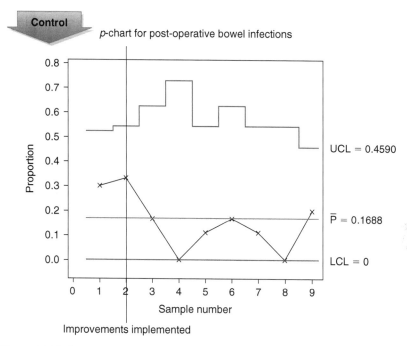

Figure 7.11 Control chart.

Cost analysis

- Emailed all patient accounts to finance.
- Finance researched cost, charges, reimbursement, and insurance group for this set of patients.
- Spreadsheet separated by patients who had developed an infection and those that did not.
- Calculated average cost, average charges, average reimbursement, and average profit for each set of patients.
- Based on these calculations, the financial impact per defect = $263.87 (20 defects for 6 months).
- This means for each defect it cost the hospital this amount.

Figure 7.12 Cost analysis (Part 1).

to achieve a 6% infection rate or lower. To that end, another control was implemented in August 2004. Now all pre-operative bowel surgery patients are provided with a Hibiclens scrub sponge and are educated concerning a pre-operative bath/shower with the antimicrobial sponge prior to arrival to the hospital. Originally, this step had been left to the surgeon's discretion. Another control has also been added that was not part of the original study, clipping of

Control   Worked with finance to calculate.

Cost analysis

- The annualized cost of a defect was $10,554. This project resulted in a 64% improvement. The annualized project savings are $6755. These numbers have been approved by finance.
- All of the infections identified during the study period were superficial infections.
- Deep and organ space infections do occur and would have a higher financial impact since they would require more intensive treatment and care.
- More private payers are moving away from per diem-driven reimbursement toward DRG reimbursement.
- This will make a greater financial impact on defects.
- The patient satisfaction piece is immeasurable. However, there is significant cost associated with satisfied customers but is difficult to quantify.

Figure 7.13 Cost analysis.

hair for pre-operative hair removal rather than shaving. This control was not included in the original project because surgeons were not utilizing clippers; therefore, there was no variable to measure. Ongoing monitoring continues as well as ongoing feedback is provided to anesthesia, the general surgeons, PACU staff, the in-patient surgical units, and out-patient department.

7.8 Conclusion

Six Sigma's future in healthcare is far from bleak. As consumers become better informed, healthcare providers must commit to achieve excellence in every aspect of the care delivered. The success stories are rapidly growing, all touting the impact of this rigorous approach to problem solving. The advantage Six Sigma brings with proven strategies allows other healthcare organizations to build on prior successes, sharing information and knowledge to ensure all patients' safety and delivering a healthcare experience defined by excellence.

Since CHC was the first healthcare organization to adopt Six Sigma as an organization-wide approach to quality improvement, the organization serves as a pioneer who led the way for many others to follow. CHC continues to move forward to constantly improve the way work is accomplished. In the words of author Ben Sweetland, 'Success is a journey, not a destination.' CHC continues on this journey to perfect processes and services utilizing the most valuable asset, its people.

8

Six Sigma within Doosan Heavy Industries & Construction Company

Sung H. Park

8.1 Introduction

Six Sigma was first introduced in 1996 to LG Electronics and Samsung SDI in Korea. Since then many world-class companies in Korea, such as Samsung Electronics, Hyundai Automobile, POSCO and Korea Heavy Industries and Construction Company, have adopted Six Sigma as their management strategy, and they have been quite successful in Six Sigma implementation. Korea Heavy Industries and Construction Company changed its name to Doosan Heavy Industries & Construction Company in 2001. It will be referred to as 'Doosan' from now on.

Six Sigma is the favorite management innovation strategy in Korea now, and more and more companies are becoming interested in Six Sigma. It is believed that Six Sigma is one of the reasons why some companies in Korea have become world-class companies and have a competitive edge. The National Quality Prize for Six Sigma was introduced in Korea in 2000. The first recipients of the prize were LG Electronics and Samsung SDI. Their Six Sigma stories can be found in the books of LG Electronics (2000) and Samsung SDI (2000).

The following improvement project was a case study at Doosan on the reduction of short shelf life material (SSLM). This was a typical non-manufacturing application which helped to develop an efficient computerized control system. The DMAIC (Define–Measure–Analyze–Improve–Control) methodology was applied in the project. Doosan learned Six Sigma management skills from General Electric in 1997, and started Six Sigma to achieve management innovation. In early 2002, the company published a book called *Six Sigma Best Practices* in which 15 Six Sigma project activities were published. The long-term vision of the company is to become a 'competitive world-class company of 21st century with the best quality and technology.' To achieve this vision, they made four strategies which were *global company*, *best quality*, *best technology* and *competitiveness*. In order to realize these strategies, they made their own management action plans (MAP), with which critical success themes (CSTs) were selected for quality and productivity innovation.

This case study presented here is one of the CSTs which is contained in the *Six Sigma Best Practices*. The Engineering and Technology Division of this company desired to solve one CST, namely 'Reduction of *Short Shelf Life Material* (SSLM).' They formed a project team with a full-time Black Belt (BB) and five part-time Green Belts (GBs) to tackle this project.

8.2 Background to Doosan Heavy Industries & Construction Company

Since its establishment in 1962, Doosan has been playing a major role in the development of Korea's national economy by supplying industrial products to domestic and overseas markets. Doosan's field of business ranges from casting and forging to nuclear and hydro and thermal power plant construction, and the company is also deeply involved in the desalination plant, environmental and material handling sectors. In fact, Doosan is the world's top manufacturer of desalination plants with a boasted 25% international market share in this area.

As the only Korean firm that specializes in power plants, Doosan is fully equipped with comprehensive production and supply systems, and manufactures everything from base material to the finished product. In particular, Doosan is busy strengthening its international competitiveness in the field of power generation by developing made-in-Korea nuclear steam generators, reactor vessels and other nuclear power-related systems. Doosan is also fully equipped with the practical and technological capability to build roads, harbors, airports and electric railroads in comprehensive manner. There are five business groups under Doosan. They are nuclear power plants, thermal power plants, turbines and generators, desalination plants, and castings and forgings.

Six Sigma is the management innovation strategy in Doosan, and under this strategy, it has four principles as follows:

1. Customers are our teachers.
2. Quality is our pride.
3. Innovation is our life.
4. People are the most important asset.

All activities including Six Sigma projects in Doosan are based on this four management philosophy.

8.3 Doosan's Six Sigma framework

Management strategies, such as Total Quality Control (TQC), Total Quality Management (TQM) and Six Sigma, are distinguished from each other by their underlying rationale and framework. Doosan adopted TQC and TQM before as most other companies in Korea did. However, Doosan now adopts Six Sigma as its major management strategy. As far as Doosan's corporate framework of Six Sigma is concerned, it embodies the five elements of top-level management commitment, stakeholder involvement, training schemes, project team activities and measurement system.

Stakeholders include employees, owners, suppliers and customers. At the core of the framework is a formalized improvement strategy with the following five steps: Define, Measure, Analyze, Improve and Control, that is DMAIC. The improvement strategy is based on training schemes, project team activities and measurement system. Top-level management commitment and stakeholder involvement are all inclusive in the framework. Without these two, the improvement strategy functions poorly. All five elements support the improvement strategy and improvement project teams.

8.3.1 Top-level management commitment

Launching Six Sigma in a company is a strategic management decision that needs to be initiated by the top-level management. All the elements of the framework, as well as the formalized improvement strategy, need top-level management commitment for successful execution. Especially, without a strong commitment on the part of the top-level management, the training program and project team activities are seldom successful. Although not directly active in the day-to-day improvement projects, the role of top-level management as leaders, project sponsors and advocates is crucial. Pragmatic management is required, not just lip service, as the top-level management commits itself and the company to drive the initiative for several years and into every corner of the company. Doosan is a well-known company in its excellent top-level management commitment for Six Sigma.

8.3.2 Stakeholder involvement

Stakeholder involvement means that the hearts and minds of employees, suppliers, customers, owners and even society should be involved in the improvement methodology of Six Sigma for a company. In order to meet the goal set for improvements in process performance and to complete improvement projects of a Six Sigma initiative, top-level management commitment is simply not enough. The company needs active support and direct involvement from stakeholders.

Employees in a company constitute the most important group of stakeholders. They carry out the majority of improvement projects and must be actively involved. The Six Sigma management is built to ensure this involvement through various practices, such as training courses, project team activities and evaluation of process performance.

Suppliers also need to be involved in a Six Sigma initiative. A Six Sigma company usually encourages its key suppliers to have their own Six Sigma programs. To support suppliers, it is common for Six Sigma companies to have suppliers sharing their performance data for the products purchased and to offer them participation at in-house training courses in Six Sigma. It is also common for Six Sigma companies to help small suppliers financially in pursuing Six Sigma programs by inviting them to share their experiences together in report sessions of project team activities. The reason for this type of involvement is to have the variation in the suppliers' products transferred to the company's processes so that most of the process improvement projects carried out on suppliers' processes would result in improvement of the performance.

Customers play key roles in a Six Sigma initiative. Customer satisfaction is one of the major objectives for a Six Sigma company. Customers should be involved in specific activities, such as identifying the critical-to-customer (CTC) characteristics of the products and processes. CTC is a subset of CTQ (critical-to-quality) from the viewpoint of the customers. Having identified the CTC requirements, the customers are also asked to specify the desired value of the characteristic, that is the target value and the definition of a defect for the characteristic, or the specification limits. This vital information is utilized in Six Sigma as a basis for measuring the performance of processes. In particular, the research and development (R&D) part of a company should know the CTC requirements and should listen to the voice of customers (VOC) in order to reflect the VOC in developing new products.

8.3.3 Training scheme and project team activities

In any Six Sigma program, a comprehensive knowledge of process performance, improvement methodology, statistical tools, process of project team activities, deployment of customer requirements and other facets is needed for its success. This knowledge can be cascaded throughout the organization and become the shared knowledge of all employees only through a proper training scheme.

In Doosan, there are five different training courses in Six Sigma. To denote these courses, Six Sigma companies have adopted the belt rank system from martial arts which is shown in Figure 8.1. These are the White Belts (WBs), GBs, BBs, Master BBs (MBBs) and Champions.

The WB course gives a basic introduction to Six Sigma. Typically, it is a 2- to 3-day course and is offered to all employees. It covers a general introduction to

Course levels		Belts
Overall vision	☺	Champion
Most comprehensive	☺☺☺	MBB
Comprehensive	☺☺☺☺☺	BB
Median	☺☺☺☺☺☺ ☺☺☺☺☺☺☺	GB
Basic	☺☺☺☺☺☺☺☺☺ ☺☺☺☺☺☺☺☺☺☺	WB

Figure 8.1 Course levels and belts for Six Sigma training scheme.

Six Sigma, framework, structure of project teams and statistical thinking. The GB course is a median course in content and the participants also learn to apply the formalized improvement methodology in a real project. It is usually a 2-week course, and is offered to foremen and middle management. The BB course is comprehensive and advanced, and aims at creating full-time improvement project leaders. BBs are the experts of Six Sigma, and they are the core group in leading the Six Sigma program. The duration of a BB course is around 4–6 months with about 20 days of study seminars. In-between the seminar blocks, the participants are required to carry out improvement projects using the DMAIC methodology. The BB candidates are selected from the very best young leaders in the organization.

An MBB has BB qualifications and is selected from BBs who have more experience of project activities. An MBB course is most comprehensive as it requires the same BB training, and additionally planning and leadership training. Champions are drivers, advocates and experienced sources of knowledge on Six Sigma. These people are selected among the most senior executives of the organization. A Champion course is usually a 3- to 4-day course, and it concentrates on how to guide the overall Six Sigma program, how to select good improvement projects and how to evaluate the results of improvement efforts.

8.3.4 Measurement system

Doosan provides a pragmatic system for measuring performance of processes using a sigma quality level, parts per million (PPM) or defects per million opportunities (DPMO). The measurement system reveals poor process performance and provides early indications of problems to come. There are two types of characteristics, continuous and discrete. Both types can be included in the measurement system. Continuous characteristics may take any measured value on a continuous scale, which provides continuous data. In continuous data, normally the means and variances of the CTQ characteristics are measured for the processes and products. From the mean and variance, the sigma levels and process capability indices can be calculated.

8.3.5 How to select project themes in Doosan?

In Doosan, as explained in Figure 8.2, the project themes are selected essentially by a top-down approach, and company CTQs are nominated as themes most of the time. The deployment method in order to select project themes is shown in Figure 8.2.

For example, let us assume that one of the company's management goals is to improve production capability without further investment. For this particular goal, each division must have its own CTQs. Also assume that the manufacturing division has such CTQs as machine down-time and rolled throughput yield (RTY). For the machine down-time, there may be more than two sub-CTQs: heating machine down-time, cooling machine down-time and pump down-time.

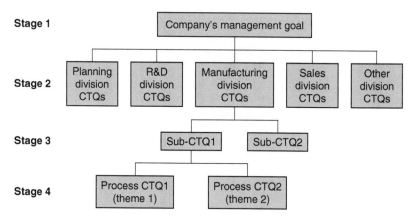

Stage 1

Stage 2

Stage 3

Stage 4

Figure 8.2 Deployment for selection of project themes.

For the sub-CTQ of heating machine down-time, process CTQ1 (theme 1) could be 'reduction of heating machine down-time from 10 h/month to 5 h/month,' and process CTQ2 (theme 2) could be '10% improvement of heating process capability.'

8.3.6 Doosan's BB course

Depending on each company, the content and duration of a BB course could be different. Doosan takes four 5-day sessions and one final graduation day. The duration is usually 4 months: 1 week for one session and 3 weeks for the practice period in each month. Hence, it takes 4 months. Usually a project is carried out during the 4-month period, and a certified examination is conducted before graduation. Also a homework assignment is given after each session. On the final graduation day, the project is presented and the BB certification is awarded. The following are the major contents of the four sessions.

(1) First session (focus on Define and Measure in DMAIC)

- Introduction to Six Sigma: The history, definition, philosophy and major strategies of Six Sigma.
- Basic statistics: Basic descriptive statistics, PPM, DPMO, defects per total opportunity (DPO), defects per unit (DPU), continuous data, normal distribution, Z-transform.

The 7 QC tools

- Six Sigma statistics: Sigma quality level, process capability, RTY, attribute data, Poisson and binomial distributions.
- Advanced statistics: Concept of statistical estimation and hypothesis testing, *t*-test, confidence interval, *F*-test, case studies and exercises.
- Correlation and regression analysis: Theories and case studies.
- Benchmarking.
- Costs of poor quality (COPQ): Quality costs, hidden factory.

- Long-term quality management: Measure process performance and case studies.
- Homework (or project) assignment (between first and second session): Several homework exercises can be assigned to make use of the above methodologies. For example:
  1. Select a process with a chronic problem which has been awaiting a solution for a long time where a certain economic advantage is to be gained by improvement. Run a project, first using the 7 QC tools and show an economic advantage.
  2. Measure process performance of at least three different characteristics and compute the sigma quality level for each one and the combination of the three characteristics.
  3. Run a regression analysis for a process, find significant factors and suggest improvements with a cost reduction potential.

(2) Second session (focus on Analyze in DMAIC)

- Review of homework assignment.
- Understanding variation, quality and cycle time.
- Process management: Principles and process flow charts.
- Measurement of evaluation analysis.
- Introduction to design of experiments (DOE): Full factorial design and fractional factorial design.
- DOE, introduction and software: Exercises with Minitab, JMP and others.
- Quality function deployment (QFD).
- Reliability analysis: Failure mode and effects analysis (FMEA).
- Homework assignment (between second and third session):
  1. Find a process where a certain economic advantage is to be gained by improvement. Run a full factorial with two or three factors.
  2. Collect VOCs and, using QFD, find CTQs which you should handle in your process.

(3) Third session (focus on Improve in DMAIC)

- Review of homework assignments.
- DOE: Analysis of variance (ANOVA), p-value, Robust design (parameter design, tolerance design).
- Response surface design: Central composite designs, mixture designs.
- Gage repeatability and reproducibility (R&R) test.
- Six Sigma deployment.
- Six Sigma in non-manufacturing processes: Transactional Six Sigma methodologies.
- Homework assignment (between third and fourth session): Select a process with a chronic problem in CTQ deployment. Screen important factors by regression analysis, optimize the process by using a robust design or a response surface design.

(4) Fourth session (focus on Control in DMAIC)

- Review of homework assignments.
- Statistical process control (SPC).
- Design for Six Sigma (DFSS).
- BB roles: Job description of BBs.
- Six Sigma and other management strategies: The relationship of Six Sigma to ISO 9000, TQC, TQM, Enterprise Resource Planning (ERP) and other management strategies.
- Six Sigma in a global perspective.
- Group work (evening program): Why is Six Sigma necessary for our company?
- Homework assignment (between fourth and graduation): Take a project where the economic potential is at least $50,000 in annual cost reduction and complete the project.

8.4 Case study of non-manufacturing applications of DMAIC: development of an efficient computerized control system

8.4.1 Define phase

There were many materials that needed storage on a shelf for some period of time to be subsequently used to create various products. Each material had its own specified shelf life-time depending on whether it was stored in a refrigerator or not. Some frequently used materials and their specifications are listed in Table 8.1. The shelf life-time was counted from the date that it was manufactured.

However, due to poor storage conditions and other reasons, the shelf life-times became short, and they could not be used in their original conditions. Such SSLM resulted in high COPQ, environmental pollution and additional test expenses.

Table 8.1 Stored materials and their specified shelf life-times

Name of material	Storage in a refrigerator		Storage in a storeroom	
	Shelf life-time (months)	Storage condition (°C)	Shelf life-time (months)	Storage condition (°C)
Mica paper tape (#77865)	6	Below 7	3	Below 23
Mica paper tape (#77906)	6	2–10	3	18–32
GI yarn flat tape prepregnated	12	Below 5	3	18–32
Mica M tape (#77921)	6	2–10	2	18–32
Modified epoxy varnish	6	Below 10	2	18–32
Polyester resin-35%	12	2–10	6	18–32
Epoxy impregnated fiber cloth (#76579)	6	2–10	1	18–32
Pa-polyester resin	10	2–10	3	18–32
Pb-catalyster	10	2–10	3	18–32
Polyester component	12	2–10	3	18–32
Glass cloth and tape	12	2–10	3	Below 25
Transposition filler	12	2–10	3	18–23
...				

8.4.2 Measure phase

During the period of July–December 2001, scrap materials were found during the process of manufacturing many products. Table 8.2 shows the scrap materials for the product, stator bar and connecting ring.

The products/processes which were of concern are listed in Table 8.3 and their current process capabilities at the time are shown in the same table.

8.4.3 Analyze phase

In order to discover the sources of defects and variation, a cause-and-effect diagram was sketched by the team as shown in Figure 8.3.

In the past 6-month period, the total defect count on the materials was 244, and the Pareto diagram for the types of defects is shown in Figure 8.4.

Figure 8.4 shows that the insufficient control of SSLM in the storehouse accounted for 42.2% of the total and the unexpected change of manufacturing schedule was responsible for 39.8% of the total defects.

Table 8.2 Scrap materials in the stator bar and connecting ring

Name of material	Purchase quantity	Scrap			Cause of scrap			
		Quantity	Unit	Number of times	Change of manufacturing schedule	Earlier purchase	No control of storage	Others
Modified epoxy varnish	92	31	GL	2			2	
Epoxy impregnated fiber cloth	7890	120	SH	2	1	1		
Glass cloth and tape	958	84	RL	5	2	1	1	1
Transposition filler	4118	55	LB	1			1	
...								

Table 8.3 Current process capabilities

Product/ process	Defect	Unit	Opportunity	Total opportunity	DPU	DPO	DPMO	Process capability (sigma level)
Stator bar and connecting ring	61	12	50	600	5.083	0.101	101,667	2.77
Stator W'g ass'y	148	13	71	923	11.385	0.160	160,347	2.49
Lower frame A.	31	14	8	112	2.214	0.277	276,786	2.09
Rotor coil A.	4	17	5	85	0.235	0.047	47,059	3.17
Total	244			1720		0.142	141,860	2.57

Defect: over shelf life-time of SSLM; Unit: four items categorized in the processing using SSLM; Opportunity: quantities of SSLM used in unit; Total opportunity = unit × opportunity; DPMO = DPO × 1,000,000; Short-term capability = Long-term capability +1.5.

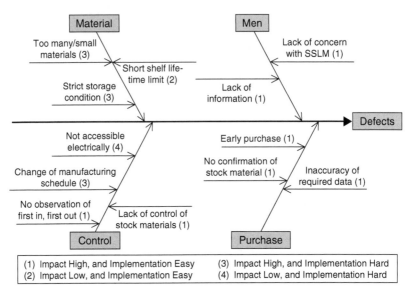

Figure 8.3 Cause-and-effect diagram for SSLM.

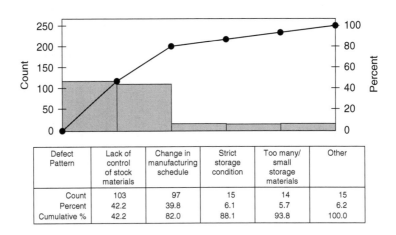

Figure 8.4 Patterns of defects.

8.4.4 Improve phase

In order to reduce the defects of SSLM, the computerized inventory control system was redesigned to increase the control efficiency of SSLM. The current process after the redesign looks as follows.

In this current process, there is no tool for checking and monitoring SSLM, and no one is assigned for checking the defects. The redesigned and improved

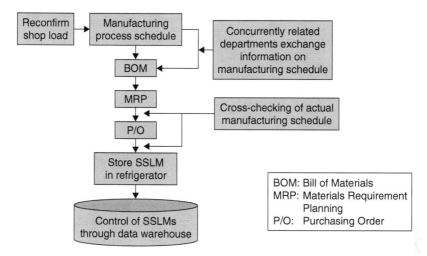

Figure 8.5 Redesigned process for SSLMs.

Table 8.4 Improved process capabilities

Product/process	Defect	Unit	Opportunity	Total opportunity	DPU	DPO	DPMO	Process capability (sigma level)
Stator bar and connecting ring	4	13	24	312	0.308	0.01282	12,820	3.73
Stator W'g ass'y	3	7	45	315	0.429	0.06667	66,670	3.00
Lower frame A.	1	6	3	18	0.167	0.05556	55,560	3.09
Rotor coil A.	0	10	5	50	0	0	0	6.00 (estimated)
Total	8			695		0.01151	11,510	3.77

process (Figure 8.5) makes cross-checking of the manufacturing schedule in advance possible. Also, the related departments can monitor and control SSLMs through the on-line system.

By practicing the improved process, the Six Sigma team obtained the following data for the first 3 months after the start of its use for SSLMs. Table 8.4 illustrates the results of the study after process improvement phase was carried out.

The quality performances of the old and newly improved processes were compared as follows, clearly showing the impact of the Six Sigma team activities.

	Before improvement	After improvement
DPMO	141,860	11,510
Sigma level	2.57	3.77
COPQ	$190,000/year	$15,400/year (estimated)

The estimated savings by this project is about $174,600/year. However, the impact was bigger on the management side rather than the financial side.

8.4.5 Control phase

In order to maintain the benefits, the team decided to follow the following control procedures.

- Update the SSLM instruction manual, and check the manual every 6 months.
- Educate the workers on SSLM information every month.
- Monitor-related data through the on-line computer system every other month.

8.5 Doosan's future plans in Six Sigma

8.5.1 Lean manufacturing and Six Sigma

Currently, there are two premier approaches to improving manufacturing operations. One is lean manufacturing (hereinafter referred to as 'Lean') and the other is Six Sigma. Since Lean has many attractive properties, Doosan has already introduced Lean for its management strategy. Doosan plans to pursue Lean and Six Sigma at the same time, and it plans to call such strategy as 'Lean Six Sigma of Doosan.'

Lean evaluates the entire operation of a factory and restructures the manufacturing method to reduce wasteful activities like waiting, transportation, material hand-offs, inventory and overproduction. It reduces variation associated with manufacturing routings, material handling, storage, lack of communication, batch production and so on. Six Sigma tools, on the other hand, commonly focuses on specific part numbers and processes to reduce variation. The combination of the two approaches represents a formidable opponent to variation in that it includes both layout of the factory and a focus on specific part numbers and processes.

Lean and Six Sigma are promoted as different approaches and different thought processes. Yet, on close inspection, both approaches attack the same enemy and behave like two links within a chain; that is, they are dependent of each other for success. They both battle variation, but from two different points of view. The integration of Lean and Six Sigma takes two powerful problem-solving techniques and bundles them into a powerful package. The two approaches should be viewed as complements to each other rather than as equivalents or replacements for each other.

Doosan has found that the following practices are good for using Lean concepts:

- Quick changeover techniques to reduce setup time.
- Adoption of manufacturing cells in which equipment and workstations are arranged sequentially to facilitate small-lot, continuous-flow production.
- Just-in-time (JIT) continuous-flow production techniques to reduce lot sizes, setup time and cycle time.
- JIT supplier delivery in which parts and materials are delivered to the shop floor on a frequent and as-needed basis.

8.5.2 Differences between Lean and Six Sigma

Doosan notes that there are some differences between Lean and Six Sigma as follows:

- Lean focuses on improving manufacturing operations in variation, quality and productivity. However, Six Sigma focuses not only on manufacturing operations, but also on all possible processes including R&D and service areas.
- Generally speaking, a Lean approach attacks variation differently than Six Sigma does as shown in Figures 8.6 and 8.7. Lean tackles the most common form of process noise by aligning the organization in such a way that it can begin working as a coherent whole instead of as separate units. Lean seeks to co-locate, in sequential order, all the processes required to produce a product. Instead of focusing on the part number, Lean focuses on product flow and on the operator. Setup time, machine maintenance and routing of processes are the important measures in Lean. However, Six Sigma focuses on defective rates and COPQ due to part variation and process variation based on measured data. Such concepts on differences between Lean and Six Sigma can be found in Denecke (1998), George (2002) and others.
- The data-driven nature of Six Sigma problem-solving lends itself well to lean standardization and the physical rearrangement of the factory. Lean provides

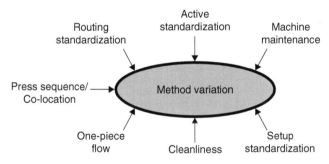

Figure 8.6 Variation as viewed by Lean.

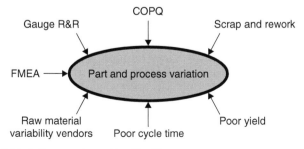

Figure 8.7 Variation as viewed by Six Sigma.

a solid foundation for Six Sigma problem-solving where the system is measured by deviation from and improvements to the standard.

- While Lean emphasizes standardization and productivity, Six Sigma can be more effective at tackling process noise and COPQ.

8.5.3 Synergy effect

The author believes that Lean and Six Sigma, working together, represent a formidable weapon in the fight against process variation. Six Sigma methodology uses problem-solving techniques to determine how systems and processes operate and how to reduce variation in processes. In a system that combines the two philosophies, Lean creates the standard and Six Sigma investigates and resolves any variation from the standard. In addition, the techniques of Six Sigma should be applied within an organization's processes to reduce defects, which can be a very important prerequisite to the success of a Lean project.

8.6 Conclusion

Doosan is now one of the best companies in Korea which employs Six Sigma as part of its management innovation initiative. Many project leaders in manufacturing and non-manufacturing areas are using DMAIC for their projects. However, in Doosan, many leaders in R&D parts are using DMADOV (Define–Measure–Analyze–Design–Optimize–Verify). It seems that both methodologies are working well in most project problem-solving.

With help from the application of Six Sigma, Doosan is emerging as a super power in heavy industries. Doosan's recent achievements include the light water reactor (LWR) project in North Korea, the nuclear power project for Qinshan of China, the Al-Shoaiba desalination plant in Saudi Arabia, the large-scale desalination plants in Al-Taweelah of the United Arab Emirates and the 2002 World Cup Stadium in Seoul, Korea.

References

Denecke, J. (1998). *Six Sigma and Lean Synergy, Allied Signal Black Belt Symposium*, AlliedSignal Inc., pp. 1–16.

Doosan Heavy Industries & Construction Company (2002). *6 Sigma Best Practices*. Doosan Heavy Industries and Construction Company, Changwon, Korea.

George, M.L. (2002). *Lean Six Sigma*, New York: McGraw-Hill.

LG Electronics (2000). *Six Sigma Case Studies for Quality Improvement*, prepared for the National Quality Prize of Six Sigma for 2000 by LG Electronics/Digital Appliance Company, Changwon, Korea.

Park, S.H. (2003). *Six Sigma for Quality and Productivity Promotion*, Asian Productivity Organization, Tokyo, Japan.

Samsung SDI (2000). *Explanation Book of the Current Status of Six Sigma*, prepared for the National Quality Prize of Six Sigma for 2000 by Samsung SDI, Suwon, Korea.

9

Application of Six Sigma in the banking industry

Alex A. Balbontin

9.1 Introduction

This chapter discusses the application of Six Sigma within the banking industry. The case for the use of Six Sigma is discussed in terms of the intrinsic nature of the banking industry and the benefits of using this business improvement/ problem-solving philosophy/methodology. The specific adoption of Six Sigma in some banks is discussed.

The author then presents a detailed example of the use of Six Sigma within JP Morgan Chase (JPMC), the leading global financial services firm. The source of the project idea and the reasoning behind its selection and prioritization is discussed. The execution of the project through the DMAIIC (Define–Measure–Analyze–Improve/Implement–Control) methodology is described. This chapter highlights the project challenges with regard to the 'hard' (data-related) and 'soft' (people-related) areas of the project. Subsequently, the case study illustrates the quantification and validation of its financial impact.

Finally, the author discusses the challenges of applying Six Sigma in banking and the future trends of the adoption of this approach in this industry.

9.2 Why Six Sigma in the banking industry?

The banking industry provides a fertile environment for the application of Six Sigma as a continuous improvement philosophy and methodology:

- It is highly transactional and volume intensive.
- Its product complexity is continuously increasing.
- Its cost of errors is very high.
- It is becoming highly competitive, demanding consolidation, and a strong expense and revenue discipline.
- In rapid growth cycles, it becomes a challenge to increase capacity while mitigating risks.
- It is just catching up with other industries in terms of the discipline and tools available for continuous improvement.

Transactional volumes are often very high, such as in the credit card services operations where some companies (e.g., JPMC, Citigroup) process millions of bills everyday; in treasury services where millions or payments for corporate and financial institutions are processed monthly; or in investor services (e.g., custody and Fund Accounting), where many clients require daily reporting on their trading activities and investment valuations (on cash and stocks). National Westminster Bank, one of the largest UK banks, handles 15-million transactions everyday for its 6-million customers, maintaining over 32.5 million direct debits/standing orders (Johnston, 1997).

Banking products are becoming highly complex and riskier, such as derivatives, or futures and options, with complex calculations linked to various markets and investment conditions (e.g., mortgage back securities, flexi-notes, etc.). In addition, one error could cost millions because of the value of transactions involved and/or the daily fluctuations of the market; for example, a bank had to pay US $10 million because of a delay on a corporate action payment due to the difference on the value of the stocks during the additional days that took to process the payment. Errors occur even in lower-complexity operations; for example, a Market and Opinion Research International (MORI) survey (1994) of 1879 retail customers found that nearly half of all customers (48%) had experienced at least one mistake during the previous 12 months. Another study conducted in 1997 found that written complaints in banks were up to 8.4% from the previous year, and bank customer satisfaction reports revealed that a quarter of all respondents found mistakes in their current accounts (Barret, 1997).

Globalization and increasingly demanding clients have affected the margin of banking institutions, requiring more efficient operations with lower expenses, forcing consolidation, off-shoring (or 'other shoring' – as most banks already operate in various continents) and outsourcing (in order to reduce expenses); and also to increase revenues; for example, one European business area of a global bank used Six Sigma in order to increase its revenue by 8% per annum through the improvement of its billing process. These pressures are greater in tight revenue cycles such as post-September 11. At the other end of the spectrum, banks have to respond quickly to market turnarounds, grasping the often short market opportunity windows without sacrificing quality or increasing their risks. In these conditions Six Sigma can be applied to increase capacity (e.g., increasing 'straight through processing' (STP)) and to build scalable/non-volume sensitive processes.

Unlike the manufacturing industry, with mature business improvement methodologies (e.g., Statistical Process Control (SPC) in the automotive industry), the banking industry has just introduced disciplined improvement methodologies in the last 10 years (initially Total Quality Management (TQM) and more lately Six Sigma). Quick-win opportunities are often available post-launch of the improvement initiatives, as banking businesses and their processes grew at a very fast pace in the 1980s to cope with the boom on the stock market, without careful planning. These processes, nowadays, have to be integrated to other organizations because of mergers and acquisitions, representing a huge improvement opportunity – also leveraging from new information technologies (e.g., by applying Six Sigma digitization).

9.3 Adoption of Six Sigma in the banking industry

General Electric (GE) Capital, the financial division of GE, launched Six Sigma in 1995 (Pande *et al.*, 2000), becoming one of the first financial institutions applying this methodology in order to increase their profitability and customer satisfaction. After this, various financial institutions and banks initially in the USA and then in Europe and Asia have followed such as American Express, Citicorp (one of the heritage companies that later became Citigroup), Bank of America, UBS, Lloyds TSB, HSBC, Zurich Financial, Bank One (George, 2003), and JPMC (Balbontin, 2002; 2004). The main driver for using Six Sigma in these companies is quite diverse, from improving service quality, to reduce expenses and increase revenues. Pande quotes various examples of the use of Six Sigma within GE Capital such as (Pande *et al.*, 2000):

- Streamlining the contract review process, leading to faster completion of deals with annual savings of US $1 million.
- Call center optimization within the Mortgage division, improving the rate of a caller reaching a *live* GE person from 76% to 99%. Beyond the much greater convenience and responsiveness to customers, the improved process had an impact of millions of US dollar in incremental revenue from new business.

Due to its high volume and value of the transactions involved, the impact of Six Sigma tools and techniques in the banking industry could be substantially higher than in the manufacturing industry (a relatively small increase on Sigma performance level could have a substantial financial impact); very often without the need of complex statistical tools. For example, the use of process maps, Pareto diagrams, and run charts to highlight opportunity expense reduction areas helped to reduce unnecessary Swift messages saving US $500,000 per annum in investor services' custody operations at JPMC. In some cases, however, more complex statistical tools are required; for example, a bank applied design of experiments to identify the most cost-effective process for finding clients with outstanding credit balances saving US $2.9 million (Doganoksoy *et al.*, 2000).

9.4 Introduction to JPMC and its use of Six Sigma

JPMC has operations in more than 50 countries, with more than 160,000 employees serving 90-million customers worldwide including the world's prominent corporate, institutional, and government clients. The firm has six lines of business: investment banking, asset and wealth management, treasury and securities services, retail financial services, commercial banking, and card services. JPMC is the product of various mergers including JP Morgan, Chase Manhattan Bank, Bank One, Chemical Bank, Manufacturers Hanover, etc.

Although heritage Chase Manhattan already had launched a TQM initiative in the mid-1990s, JPMC inherited its Six Sigma initiative from heritage JP Morgan, where it was launched in 1998 and from Bank One, where it was launched in 2002 (George, 2003). Heritage JPMC had a global and cross-business, mature Six Sigma program encompassing a DMAIIC and Design for Six Sigma (DFSS) methodologies with business aligned support teams distributed across the globe. Heritage Bank One, however, did not have a cross-business initiative and it was mostly implemented across the National Enterprise Operations Group (George, 2003) where the Focus 2 program was launched in early 2002 integrating Six Sigma and Lean methods (following the introduction of Focus 1 in the 1990s as a simple problem-solving approach to address gaps – alike the introduction of the 'Workout' program in GE). Since our JPMC–Bank One merger, Six Sigma has evolved within the bank. Our management has decided to apply a *light* approach toward the use of Six Sigma in our new company, reducing the resources used to support the program (moving more toward the Bank One model). Its application is still focused on improving client satisfaction, reducing risks, increasing revenues, and reducing expenses. Some examples of JPMC projects on each of these benefit areas are:

- *Client satisfaction*: Account set-up time reduction, reduction of client enquiries, reduction of referrals, reduction of compensation claims, Management Information System (MIS) client reporting optimization (most of them with impact on expense reduction), etc.
- *Risk reduction*: Selective introduction of controls (through the use of process mapping, failure modes and effects analysis and error-proofing techniques within the projects) in high-risk processes such as corporate actions, derivatives and options processing, etc.
- *Expense reduction*: Increase of STP to reduce manual intervention, Swift messaging cost reduction, functionalization and hubbing of back- and middle-office processes (with later move to low-cost locations), consolidation of common functions, technology infrastructure cost reduction (e.g., server storage and mobile phone cost reduction), etc.
- *Revenue increase*: Billing process optimization, cash investment products revenue increase, cross-selling optimization, etc.

In many cases we apply the whole Six Sigma methodology, but very often, we only use some of the tools to achieve our objectives. As Jamie Dimon, current President and Chief Operating Officer of JPMC (to succeed William B. Harrison as CEO in 2006), who was just named '2004 Banker of the Year' by American Banker, has stated (in a business town-hall meeting in New York in March 2004):

We used Six Sigma at Bank One in a lot of our operations. It helps to build great systems, great operations, great reporting, know the facts; help people get in the room, look at the processes, improving them, and find out why we have errors.

However, he further emphasized:

> *Six Sigma is a wonderful tool if used properly. I've seen corporations doing a great job on it and I've seen corporations where it just becomes a big bureaucracy.*[1]

Therefore, we are very selective and pragmatic in our approach in order to get the right balance by using the right tools to achieve optimum results in a timely manner.

9.5 Introduction to JPMC investor services and project background

The specific project presented in this case study took place in the area of Fund Accounting which is part of Investor Services within Treasury Securities Services (TSS) in the UK. Investor Services has been recognized by *Global Investor* as the best Global Custodian Overall for 2004 and it has US $7.9 trillion in assets under custody. It has also been recognized by *The Banker* as the 'Global Securities Services House of the Year' for 2004. Investor Services offers Fund Accounting on the back of the custody service. Fund Accounting calculates and delivers the value of funds' portfolios to fund managers, based on the amount and value of their asset holdings. This project was needed in order to streamline the valuation delivery to external clients, as the business had missed its delivery deadlines as committed in its Service Level Agreements (SLAs). One key client (which generated US $14 million in revenue in 2002) was highly unsatisfied about these missed deadlines, being a major project drive for the retention of this client. Therefore, this project was mandatory, with high management priority.

9.5.1 Define phase

In order to perform the portfolio valuation, the Fund Accounting platform receives the asset holdings' status through feeds from the custody system. Then, it incorporates the pricing feeds (for assets and foreign exchange rates (FX)) from various sources such as Bloomberg and Financial Times (FTID). The latter is a semi-automated process, requiring manual intervention (see supplier–input–process–output–customer (SIPOC) in Figure 9.1).

A high-level process cycle time analysis found that the overnight batch process took 45% of the total process time whereas the business process only took 33% (see Pareto in Figure 9.2). The batch process delays the online availability of the Fund Accounting platform taking on an average of more than 6 h

---

[1] Note that Jamie Dimon got to know Six Sigma while working in Citibank under the leadership of Sandy Weill (Langley, 2003) just after the merger between Travelers and Citicorp. Heritage Citicorp had a strong program supported by John Reed (his CEO). Citicorp executives never managed to convince Sandy Weill of the benefits of Six Sigma and James Bailey (Heritage Citicorp corporate Six Sigma quality director asked to be reassigned).

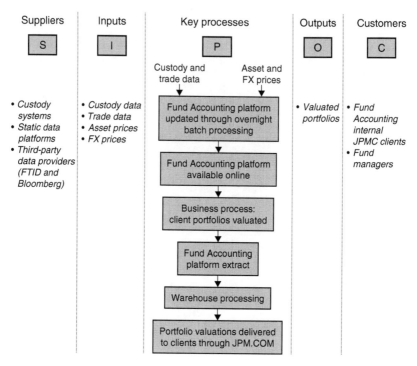

Figure 9.1 Portfolio valuation SIPOC.

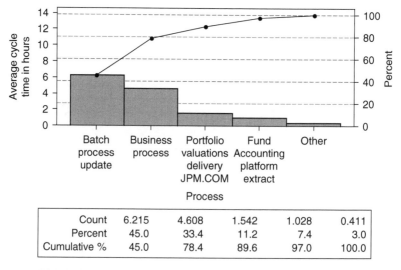

Count	6.215	4.608	1.542	1.028	0.411
Percent	45.0	33.4	11.2	7.4	3.0
Cumulative %	45.0	78.4	89.6	97.0	100.0

Figure 9.2 Pareto of cycle time distribution of portfolio valuation process.

Table 9.1 ANOVA on process cycle time hours vs. process name

Analysis of variance for cycle time

Source	DF	SS	MS	F	P
Process	4	1878.507	469.627	750.21	0.000
Error	365	228.488	0.626		
Total	369	2106.995			

				Individual 95% CIs for mean based on pooled StDev			
Level	N	Mean	StDev	----------+----------+----------+----------			
Batch	74	6.2162	0.9034				(*)
Business	74	4.6077	1.3159			(*)	
FA extract	74	1.0277	0.2661	(*)			
Warehouse	74	0.4110	0.4538	(*)			
JPM.COM	74	1.5322	0.5527	(*)			
				----------+----------+----------+----------			
Pooled StDev =		0.7912		2.0	4.0	6.0	

to be performed. Therefore, by improving this process, additional time can be available to conduct the business process so that the 11:00 a.m. delivery SLA is achieved (the technology subject matter experts working on the project estimated that 2–3 h could be gained to perform the business process). We decided to leave the business process optimization out of the scope of our project, as it was undergoing substantial re-engineering (although close communication was in place to understand relevant interdependencies).

An analysis of variance (ANOVA) analysis in which the differences between the average cycle times were tested gave us statistical confidence on the project focus. See Table 9.1 where *p*-value shows that there is a statistically significant difference between the process cycle times using a 95% confidence interval. Therefore, the project focused on making this platform available earlier. The project objective was defined as follows:

To improve the online availability delivery of the Fund Accounting Platform to the business to 06:30 a.m. by the end of June 2003 and target to improve to 06:00 a.m. by September 2003.

Based on our SLA, the client needs to receive its portfolio valuations by 11:00 a.m.; the latter SMART[2] objective aimed to ensure the business had enough time to value the asset portfolios. A one-page project charter (used in all our Six Sigma projects within JPMC) was created to capture the project objective, its problem statement, its scope, its potential impact, the team involved, and the high-level project timeline for each of the DMAIIC phases.

The close involvement of operations and technology was a critical success factor in the project execution. In parallel, as substantial variation was also found in the business process (see Table 9.1), we fed this information to its

---

[2] SMART: Specific, Measurable, Aggressive (also Achievable), Relevant, and Timely.

re-engineering team (who was in parallel looking at the workflows and sources of variation within this process).

9.5.2 Measure phase

A data collection plan was used in order to establish the key metrics to be collected in terms of inputs and outputs affecting our online availability. The voice-of-the-business (VOB) and voice-of-the-customer (VOC) conducted as part of Measure confirmed that the online availability was critical-to-quality (CTQ):

- *VOB*: Critical to UK Fund Accounting to have a consistent arrival of online availability to meet the Client SLAs. Target is to move the technology SLA to 06:00.
- *VOC*: Critical to key client to receive daily data files by 11:00 a.m. SLA consistently in order to meet business objectives. Data provided drives key client's internal systems and deliverables.
- *CTQ Metric*: Timely arrival of critical interfaces to achieve online availability technology SLA target ideally by 06:00 a.m. (current SLA is 07:00 a.m.). Also, technology SLA achieved consistently.

The measure phase of this project also focused on baselining the online availability and on mapping the batch feed process. The data was extracted from the Fund Accounting platform as it records daily the specific times when each process starts and ends. As shown in Figure 9.3, a run chart was used to baseline the online availability. The chart shows the CTQ of 07:00 a.m. online availability as the upper control limit.

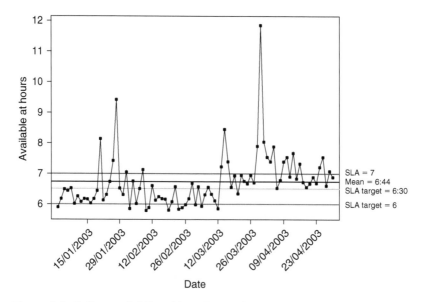

Figure 9.3 Online availability of baseline performance – run chart.

As shown in Figure 9.4, a monthly baseline Sigma was calculated from December 2002 to April 2003. In April 2003, the Sigma level was 1.5 Sigma which means that the 07:00 a.m. online availability was missed 50% of the time. The team aimed to answer why the performance of our CTQ got worse from March 2003. During that month an Income-Generator calculation process was introduced and conducted overnight.

All the key processes related to the online availability were mapped and the key batch processes were broken down into backups, interfaces, sweep, generators, and translate:

1. *Backups*: The process to ensure full recoverability of the Fund Accounting platform data prior to the start of any batch processing. So that in the event of an issue requiring recovery the days online processing is not lost.
2. *Interfaces*: The delivery of data from many other applications to build/consolidate the Fund Accounting platform to provide all the data to give current positions on the Funds.
3. *Sweep*: The first part of the process that consolidates the data from the interfaces and adds it into the Fund Accounting platform files. Through this process, we update the batch so that we can save online each fund being actioned individually the next day. This is a high-volume data processing function which must run in sequence with sweep/generators and translate.
4. *Generators*: Provide key data to the system via a batch process rather than online entry. This is a high-volume data processing function which must run in sequence with sweep/generators and translate.
5. *Translate*: Update all data files. Once completed, the system is ready to begin the online data entry and to deliver data to the clients. This is a batch function that could be delivered online but that would impact client delivery timetables. This is a high-volume data processing function which must run in sequence with sweep/generators and translate.

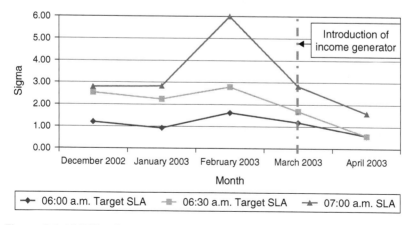

Figure 9.4 UKMT online availability Sigma performance.

9.5.3 Analyze phase

This phase focused on identifying the root causes of the delays on the Fund Accounting platform online availability. These root causes were grouped around the batch processes.

A Pareto chart based on the June 2003 missing SLAs was used to identify the key defects (see Figure 9.5). The late start of the batch and the execution of the generators were found to cause 57% of missing the 06:00 a.m. SLA. The execution of the sweep/generators was also a main defect causing 21% of delays.

Then, test for equal variances was conducted on the timings of all processes and batch sub-processes. It was found that the business process and the interface batch sub-process had a significantly greater variation than the other sub-processes (see Figure 9.6).

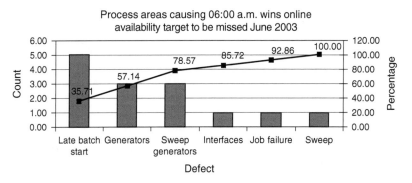

Figure 9.5 Pareto analysis to identify key defects.

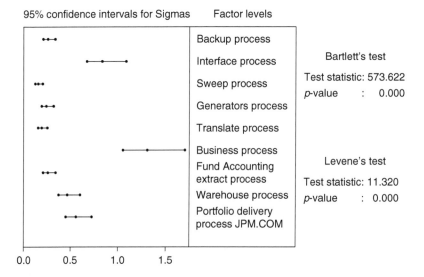

Figure 9.6 Test for equal variances – cycle time for each process component.

Then, various hypotheses were tested with regard to the root causes of such variation. Here, again, our findings were fed to the team working re-engineering the business process.

As shown in the scatter diagram presented in Figure 9.7, the growth on the number of funds significantly delayed the time to perform the sweep sub-process – a similar relationship was found for the interface sub-process. The latter was expected as various processes (including complex calculations) are performed in a batch process involving all the funds prior to the online avail-ability. Due to this finding, we had stronger focus on monitoring our pipeline of new funds coming into our process. We used our regression model in order to predict the impact of the growth of the volume of funds in the timings of the process. We also identified some alternative ways to have a more scalable sys-tem, so that we could grow without sacrificing the processing speed.

Another tool we used as part of the analysis was an Ishikawa diagram (fish-bone) in order to identify the root causes to the delay of the batch process.

9.5.4 Improve/Implement phase

Various brainstorm sessions involving operations and technology took place. The impact–complexity matrix shown in Figure 9.8 summarizes the generated ideas.

There was a strong focus on quick wins and short-term enhancements. Some of the strategic enhancements could not take place until the next release of the Fund Accounting platform (about a year from the idea generation). Key actions implemented included the multi-streaming of our translate sub-process so that

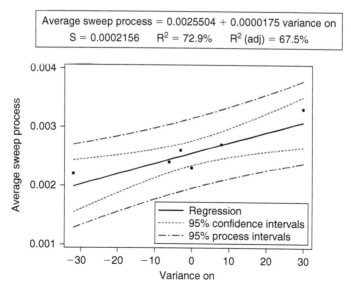

Average sweep process = 0.0025504 + 0.0000175 variance on
S = 0.0002156 R^2 = 72.9% R^2 (adj) = 67.5%

Figure 9.7 Regression analysis – timing of sweep processes vs. variance on the number of live accounts.

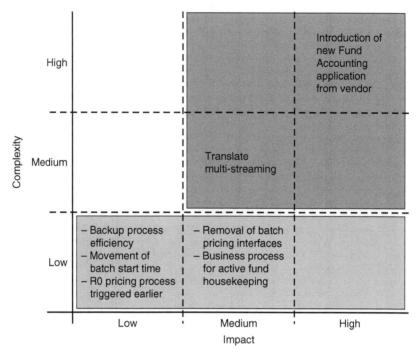

Figure 9.8 Impact–complexity matrix used to prioritize improvement ideas.

we could take additional funds without sacrificing the process speed (as funds are run through parallel processes), the removal of batch pricing interfaces, and the proactive management of the funds' pipeline (the latter two actions with small complexity and medium impact). In parallel, some business process changes were put in place.

9.5.5 Control phase and project results

As part of the on-going controls, a dashboard was introduced monitoring various metrics including inputs and outputs. As expected, online availability was monitored as the key output. The growth of funds and the time of individual sub-processes were monitored as key inputs. This dashboard is now owned by both, technology and operations, and it is updated daily. There is monthly review involving operations and technology where the dashboard is used as the basis of discussion on the performance of online availability.

From April to November 2003, process variation has been reduced, with online availability performing well under the 6.00 a.m. target SLA. During this period a 35–45% volume growth has been seen within the application. New business functionality requirements have seen a new 50-min process added to the overnight batch cycle.

As shown in Figure 9.9, the process variation was significantly reduced achieving a Six Sigma performance in January 2004 (for our toughest SLA).

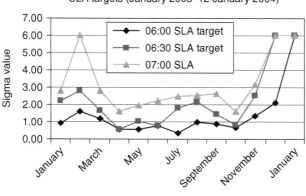

Figure 9.9 Run chart showing improved Sigma performance.

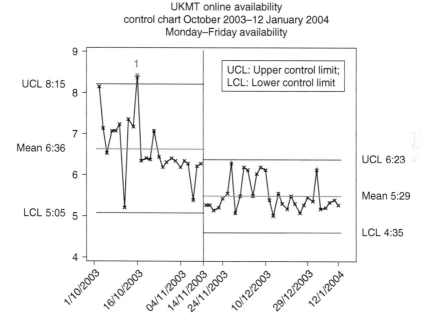

Figure 9.10 Control chart showing reduced variation and the shift of the average online availability time.

As shown in Figure 9.10, a control chart was used to statistically confirm the improvements achieved by this project. Variation was substantially reduced and our control limits were compressed and shifted down. By November 2003 the total batch process time was reduced by 50 min since June 2003 to around 5 h and 30 min with an online availability around 05:30 a.m.

The improved online availability delivery times and batch stability delivered since November 2003 has led to the negotiation of a new earlier internal

SLA of 06:00 a.m. in December 2003, from technology to operations. This will enable the project to meet the target goals with the challenge of added functionality and continuing business growth. The business process was also improved and we are now delivering the portfolio valuation to our key client (which generates US $14 million of revenue per year) well within the 11:00 a.m. SLA. This has allowed us to retain this client and it has had a positive impact across all our other clients.

In addition to the financial impact on revenue retention, the earlier online availability of our Fund Accounting platform had some spin-offs saves such as the reduction of staff overtime (approximately US $14,400 → 720 h per year → 1 overtime hour per day reduced for three clerical employees).

9.6 Conclusion

This case study illustrates an example on how we applied the Six Sigma problem-solving methodology involving operations and technology in order to retain a key client. We looked at the processes and at its relevant metrics through selective use of Six Sigma tools. The tools helped us to focus on the right areas up-front, setting our scope boundaries – we focused in making the Fund Accounting platform available earlier to Operations, increasing their processing window, while on-going re-engineering took place on the business process (as a separate initiative, with close communication with our project team).

The analysis tools helped us to identify the key process inputs and root causes delaying our Fund Accounting platform online availability. They provided the team with relevant facts, to improve the process (e.g. regression analysis allowed us to quantify the impact of the fund growth on our platform capacity). Based on this, a dashboard was created including relevant metrics – some of which were not previously monitored (e.g. fund growth and elapsed time on each sub-process component). This dashboard is critical to sustain our improvements, as the business is highly dynamic (e.g. new products are very often introduced with additional processing demands with the potential of delaying our online availability – as we found with the Income-generator process).

9.7 Some challenges of Six Sigma in the banking industry

There are various challenges when applying Six Sigma in the banking industry, such as: facing data collection issues, dealing with highly dynamic processes, and presenting the results of the analysis using the banking language (not the statistical language) in order to get support on the recommendations. Each of these challenges is discussed next.

We have lots of data available in the banking industry, however, most of the time, these data are not readily available for its analysis (e.g. we need to conduct a tailored extract from our systems to obtain the key data for our projects, we often face completeness and accuracy issues, etc.) and very often it is easier to collect the data from scratch than adjusting our existing data. Very often, we deal with discrete data where discrete repeatability and reproducibility analysis (e.g., Attribute Gage R&R) needs to be conducted to ensure good data quality. The latter is particularly true when conducting deep-dive analysis where we need to classify our errors by consensus from our employees, defining accurate error categories. For example, we have applied these techniques at JPMC when categorizing client enquiries, non-STP reasons, client referrals, etc. As we are dealing with discrete data, we require larger samples than when working with continuous data, taking longer to collect our data. An added challenge is the source of our data: unlike manufacturing, our supplier if often our customer and we have low bargaining power. Therefore, we need to demonstrate to our clients 'what is in there for them' in order to get them involved in the data collection process.

Our processes are highly dynamic and sensitive to the changes in the financial markets. Most of the time, they are seasonal and they are often non-normal and out-of-control (statistically speaking). The latter intrinsic characteristics of our banking processes must be considered when selecting our samples and analytical tools. In some cases, we have to transform the data (normalize it) or to use the applicable tools for non-normal data (e.g., X Bar-R control charts which do not rely on having normally distributed data).

Last but not the least, and a shared challenge with all other industries, we need to present our recommendations using the business language rather than the statistical language. Although we use the Champion training to raise the senior management awareness on the Six Sigma philosophy and methodology (at least at the high level), only very few managers have a statistical background which means that we always have to translate statistical findings into a common language, easy to be understood by everyone. Sharing results in terms of improvements on Sigma performance help, although we also need to translate the benefits of specific recommendations into dollar terms (e.g., through estimates on employee time saved, capacity increase, staff reduction, revenue increase, prevention of errors and financial losses, etc.). The latter is very important, as discussed earlier in this chapter; a small improvement on the Sigma level of performance could represent a substantial impact in the banking industry (because of our high cost of errors).

9.8 Future trends of Six Sigma in banking

The adoption of Six Sigma in banking as a continuous improvement (and problem-solving) philosophy and methodology will continue and grow in the coming years (see Hoerl, 2004). Six Sigma will be, however, one tool available in the toolkit which is being enriched with additional tools such as Lean and Hoshin Kanri (we have seen these in various banks such as Bank of America, UBS, etc.). As the use of Six Sigma extends and matures in our industry, it will be easier to share a common project approach and language when working with customers and suppliers. At JPMC we have a few projects with our clients where their familiarity with Six Sigma techniques has substantially helped to communicate with them and to speed the project execution.

Although most of the application of Six Sigma in banking has been in the back- and middle-office operations, we have had various success stories in the front office at JPMC (as other banks had – e.g., Bank of America). We should expect an increase on its usage on this space. The same is true in the area of banking technology, where we have experienced various successes in the infrastructure and technology servicing areas. The area of Application Development (software development), however, represents a future opportunity (relying on having Six Sigma professionals with strong software development expertise). We have to consider that most of our banking process improvement projects rely on new or updated technology to support them. Although we have made some progress at JPMC on the introduction of a Digitization methodology (previously applied at GE), this is still an area with further opportunities that will enable us to leverage from Internet technologies (requiring the use of DFSS).

We have mostly used DMAIIC at JPMC (I would say that about 90% of our projects applied this methodology). However, DFSS has also proved useful when designing new banking processes; for example, a Quality Function Deployment was used in order to determine specific process requirements based on client CTQ requirements when restructuring the middle office (client services) in investment banking futures and options. There are opportunities for greater use of DFSS as Six Sigma becomes more mature in the banking industry, as in many cases incremental improvement is not good enough and we need to radically design new processes (and/or new products).

References

Balbontin, A. (2002). *Deploying Six Sigma at JP Morgan Chase*. Six Sigma Conference, April, Cranfield University, Cranfield, UK.

Balbontin, A. (2004). *Deploying Six Sigma in Investment Banking – The JP Morgan Experience*. Customer-Focused Six Sigma Forum, February, London, UK.

Barret, P. (1997). Banks lend an ear to service: improved customer service. *Marketing*, 16 January, 16–20.

Doganoksoy, N., Hahn, G. and Hoerl, R. (2000). The evolution of Six Sigma. *Quality Engineering*, 12(3), 313–326.

George, M.L. (2003). *Lean Six Sigma for Service*. USA: McGraw Hill.

Hoerl, R. (2004). One perspective on the future of Six Sigma. *International Journal of Six Sigma and Competitive Advantage*, 1(1), 112–119.

Johnston, R. (1997). Identifying the critical determinants of service quality in retail banking: importance and effect. *International Journal of Bank Marketing*, 15(4), 111–116.

Langley, M. (2003). *Tearing Down the Walls*. USA: The Wall Street Journal Books, USA.

MORI (1994). Satisfaction with Bank and Building Society Services. London, UK: Research conducted for the British Bankers Association, Summer.

Pande, P., Neuman, R. and Cavanagh, R. (2000). *The Six Sigma Way*. USA: McGraw Hill, NY, USA.

Sureshchandar, G.S., Rajendran, C. and Anantharaman, R.N. (2003). Customer perceptions of service quality in the banking sector of a developing economy: a critical analysis. *International Journal of Bank Marketing*, 21(5), 233–242.

10

An application of Six Sigma approach to reduce fall hazards among cargo handlers working on top of cargo containers

T.Y. Ng, F. Tsung, R.H.Y. So, T.S. Li and K.Y. Lam

10.1 Introduction and background

This chapter reports a study to enhance the level of occupational safety among mid-stream cargo operators located at the public cargo working areas (PCWAs) in Hong Kong by controlling all appropriate factors and reducing operation errors using the Six Sigma DMAIC (Define–Measure–Analyze–Improve–Control (DMAIC) methodology).

Hong Kong has been one of the world's busiest container ports with an average annual throughput of 18 million twenty-foot equivalent units (TEUs). In 2002 alone, Hong Kong's container throughput reached 19.1 million TEUs. There are three main modes of port operations in Hong Kong: (i) container terminals, (ii) mid-stream, and (iii) river trade. The container terminals mainly handle cargo containers transported by large ocean-going vessels which account for about 62% of the total port throughput (Hong Kong Special Administrative Region (HKSAR), 2003). The river trade terminal mainly handles the cargos transported by river trade vessels traveling on the Pearl River to-and-from Guangdong, China, and it handles about 21% of the total port throughput (HKSAR, 2002). The rest of the sea freight is handled by various mid-stream operators whose average cargo throughput is about 17 million tonnes per annum and shares 17% of the total port throughput (HKSAR, 2003). Within this share of 17%, about 3% of the cargos (i.e., about 540,000 TEUs) were handled at the PCWAs.

10.2 Problem statement

From 1992 to 2002, 50% of the fatal accidents in the cargo handling industry was because of people falling from height (HKSAR, 2003). This translates to an average death toll of four workers per year and all of these fatal accidents occurred in mid-stream port operations. Hence, reducing the number of accidents due to fall among the workers in mid-stream operations was the targeted problem in this study.

10.2.1 Details of cargo handling activities in mid-stream operations

Since the mid-stream port operations were responsible for the fatal accident in the cargo handling industry, its operation was analyzed. Mid-stream operations involved loading and unloading of cargos (i) between ocean-going container vessels moored at buoys and barges as well as (ii) between the barges and the PCWAs. The former involved Chinese seamen without Hong Kong citizenships and the latter involve workers with Hong Kong citizenships. This study focuses on the latter because there is no official record of accidents for the former.

A barge is a flat-bottomed vessel equipped with a derrick crane to transport containers through rivers. At PCWAs, the loading and unloading of containers are carried out using the simple derrick cranes located in the barges. Figure 10.1 shows a typical work flow diagram for container loading activity in a PCWA. The use of the simple derrick cranes required the workers to work on top of the containers to hook and unhook the slings of the derrick cranes to-and-from the containers.

10.2.2 Objectives and overview

The objective of this study was to use Six Sigma methodology to reduce the occurrence of fall accidents at the PCWAs in Hong Kong. The five phases of the Six Sigma methodology were applied: Define, Measure, Analyze, Improve, and Control (DMAIC).

In the define phase, the significance of fall accidents at the PCWAs was identified through past statistics and survey. Factors affecting near-misses and hazardous levels of different procedures were measured through questionnaire survey during the measure phase. In the analyze phase, the factors related to fall accidents were validated and critical factors were identified. Finally, during the improve phase and control phase, recommendations were developed to reduce the critical factors and plans to maintain the performance were established.

10.3 Define phase

10.3.1 Past statistics on fall accidents in the cargo handling industry in Hong Kong

From the past accident statistics obtained from HKSAR Marine Industrial Accident Statistical Reports between 1992 and 2002 (Figure 10.2), there were, on average, 63 accidents related to fall of a person per year in the cargo

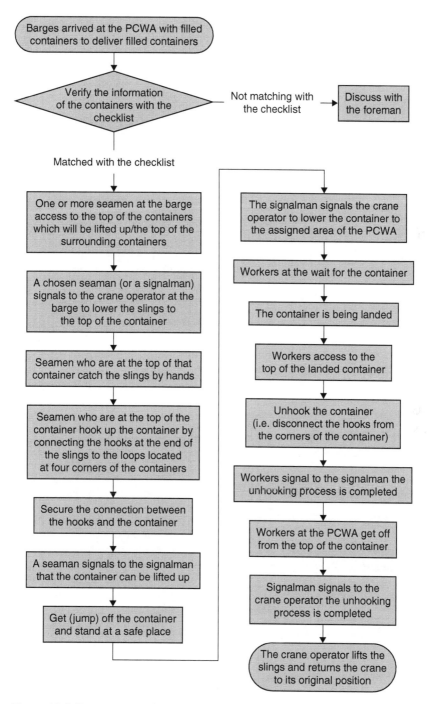

Figure 10.1 Process map of a typical container handling operation between a barge and a PCWA.

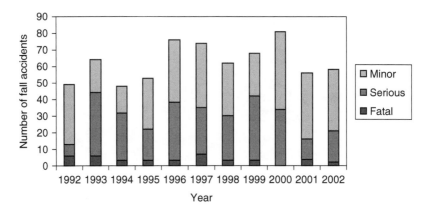

Figure 10.2 The statistics of the accidents related to the fall of a person happened in the cargo handling industry in HKSAR (1992–2002), the accidents were categorized into fatal, serious, and minor. The statistics were obtained from HKSAR Marine Department (HKSAR, 2003).

handling industry resulting in about four deaths per year. On average, there were estimated 520,000 container loading operations in the PCWAs per year. Based on this estimation, the sigma level of the fall accidents was calculated to be ranged between 4.5 and 4.7. This implies that there were about 1350 fall accidents per million operations. Since Hong Kong is among the most busy logistics cargo port in the world, maintaining a safe and efficient level of cargo handling services is of utmost important. In 2002, the Occupational Safety and Health Council of Hong Kong had highlighted the reduction of fall accident with hooking-and-dehooking cargo containers as one of the priority concerns (HKSAR, 2002).

10.3.2 Interview surveys to establish the working procedures

Interview surveys were conducted with eight experienced workers who had accumulated over 50 years of service. The purpose of the interview was to obtain accurate account of the working procedures at the PCWAs. In particular, details on all procedures that were considered to be hazardous were collected. Figure 10.1 shows a typical work flow of transporting a container from a barge to a PCWA. Inspections of Figure 10.1 indicate that the hooking-and-dehooking procedures expose the worker to fall hazard – a concern shared by the Hong Kong Occupational Safety and Health Council (HKSAR, 2002).

10.4 Measure phase

10.4.1 Preliminary interview

The eight experienced workers interviewed in the define phase were invited to list out all the factors that are associated with work safety in PCWAs. Results indicated that 38% of the responses have to do with long working hours due to (i) irregular arrival times of cargo vessels and (ii) the reduction in number of co-workers because of the on-going price wars within the container handling industry. About 22% of the responses were related to the disregard of safety guidelines suggested by the Hong Kong Occupational Safety and Health Council. The given reasons for the workers to disregard the safety regulations were mainly because of inconvenience over the use of the safety appliances. For example, wearing safety shoes, as suggested by the guidelines, was disregarded because the safety shoes make it very difficult for workers to climb to the top of the containers. Also, the use of safety harness, as suggested by the guideline, was disregarded because workers believed that they could not find a suitable anchor point to apply the safety harness. In other words, workers considered the use of safety appliances as not practical. Another given reason was that the cargo service companies did not provide suitable safety appliances such as elevated fenced platforms for workers to get to the top of the containers. About 16% of the responses were associated with strong wind and wave during bad weather conditions. Strong wind could cause the slings of the crane to swing and knock a worker off a container. This information provided the basis for a larger scale questionnaire survey conducted to identify critical factors and procedures related to fall hazards in PCWAs.

10.4.2 Measurement tree

Based on the information collected from the eight experienced workers, a measurement tree was constructed according to the Six Sigma strategy. Inspections of Figure 10.3 indicate three main types of measurements: (i) hazardous procedures that can cause workers to fall from the top of the containers, (ii) factors affecting the occurrence of fall accidents, and (iii) near-misses in PCWAs related to fall accidents. Data based on expert opinion collected from experience workers accumulated over 50 years of cargo handling service at PCWAs. This tree organized the expert information in a systematic way.

10.4.3 Questionnaire development

To further substantiate the expert opinion, two major questionnaire surveys were conducted. The questionnaire was developed based upon the information obtained from the interviews conducted in the define phase and through thorough discussions with a practicing safety engineer (Mr. T.S. Li – a co-author) and council members from the Logistics Cargo Supervisors Association (LCSA) (one council member's family has been handling cargo at the PCWAs for three generations). A pilot survey was conducted on 42 mid-stream operation workers at one of the PCWAs. The initial findings were presented in the Safety and Health Expo 2003 (Ng et al., 2003) and to the members of the research committee of the HKSAR Occupational Safety and Health Council. Using the feedback and comments obtained from representatives of the LCSA, Labour

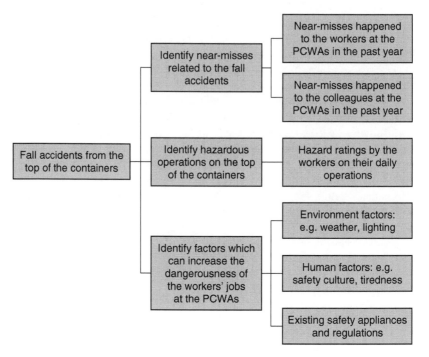

Figure 10.3 A measurement tree constructed to identify the three types of measurements needed in a questionnaire survey among the cargo handling workers at the PCWAs in Hong Kong.

Department, Occupational Safety and Health Council, and attendee of the Safety and Health Expo, the initial questionnaire was revised. This revised set of questionnaire was then used in a larger scale survey covering six out of the eight PCWAs in Hong Kong. These PCWAs included New Yau Ma Tei PCWA, Rambler Channel PCWA, Cha Kwo Ling PCWA, Wan Chai PCWA (later Chai Wan PCWA, HKSAR, 2003), Kwun Tong PCWA, and Stonecutters Island PCWA. The other two PCWAs that were not included in this survey were Tuen Mun PCWA and Western PCWA. Workers in Tuen Mun PCWA had already participated in the preliminary questionnaire survey and they did not want to participate again in the main questionnaire survey. Furthermore, the goods handled at the Western PCWA were mostly not containerized, therefore Western PCWA was not included in the main questionnaire survey. In other words, nearly 100% of the container handling workers in PCWAs took part in the questionnaire surveys.

10.4.4 The reliability of the questionnaire

The test–retest method was used in this study to examine the reliability of the questionnaire data (Litwin, 1995). Workers were requested to complete the same set of questionnaire twice within a few hours gap in between. A longer

Table 10.1 The three most hazardous procedures while working on the top of the containers in the PCWAs

Hazardous procedures	Proportions of workers rated the procedure as most hazardous (each worker can vote for more than one procedure)
Jumping from the containers which were still being carried in the mid-air	25/131 (19.1%)
Taking rest on top of the containers	23/131 (17.6%)
Standing on top of the containers which were being carried in the mid-air	21/131 (16.0%)

time gap was not used as it would greatly reduce the response rates in our study. The two completed questionnaires from the same worker were used to calculate the correlation coefficient indicating the reliability of the questionnaire. The order of the questions was rearranged to be different between the first and the second set to avoid copying using memory. This questionnaire survey did not record the names of the workers. One hundred and thirty-one workers participated in this questionnaire survey and 99 of these workers were willing to complete two sets of questionnaire. Answers to each questionnaire were coded into 236 items before conducting the correlation tests. Fifty-seven percent of the items reported in the first and second questionnaire were significantly correlated with correlation coefficients that exceeded 0.7. These highly correlated items include (i) possible factors for fall accident, (ii) hazardous operations, and (iii) examples of near-misses in the PCWAs. Examination of those items with correlation coefficients below 0.7 revealed two possible reasons for the differences: (i) the impatience of workers as well as (ii) sudden physical presence of their employers while they were filling in their questionnaires. The former was based on the larger portion of unanswered questions in the second attempts and the latter was based on matching between the circumstantial evidence observed by the research staff. Items with correlation coefficients below 0.7 were excluded in the subsequent analyses.

10.4.5 Measurement I: The hazardous procedures related to working on top of the containers in the PCWAs

The three most hazardous work procedures related to fall accidents were identified to be (i) jumping from containers to the ground when the containers were being carried by the slings and situated in the mid-air, (ii) taking rest on top of the containers, and (iii) standing on top of the containers which were being carried by the slings and situated in the mid-air (Table 10.1). To our surprise, the hooking-and-dehooking procedures were not frequently reported as highly hazardous, although these two procedures were identified as the most hazardous procedures in the initial interview and were the primary concerns of the Hong Kong Occupational Safety and Health Council. There were only 9 out of 131 workers (i.e., 7%) who agreed that hooking was the most hazardous procedure and 6 out of 131 (i.e., 4.5%) workers agreed that dehooking was the most hazardous procedure.

10.4.6 Measurement II: The factors which are highly related to fall accidents from the top of the containers in the PCWAs

Table 10.2 shows the nine most frequently reported factors for falls from the top of the containers. These factors were grouped into three clusters: (i) personal factors, (ii) environment factors, and (iii) disregard of safety regulations. Personal factors include not concentrating at work (31.3%), fail to follow proper work procedures (22.1%), drinking alcohol (20.6%), and tiredness (18.3%). Environment factors include the use of worn-out hooks and slings (42.0%), oil on shoes' soles (27.4%), and swinging slings (19.8%). Examples of disregard of safety regulations include the disregard of regulations (21.4%) and failure in using proper safety appliances (19.8%). The percentages indicate the proportions of 131 workers ranked that particular factor as the most impor-tant one. Since workers can select more than one most important factor, the percentages do not add up to 100%.

The disregards of safety appliances were further investigated. There were eight types of safety appliance mentioned in the safety guidebook to be used in cargo handling industry (HKSAR, 1999). One of these safety appliances was the safety harness and its main purpose was to prevent fall and it was the only safety appliance mentioned to prevent fall accidents. Inspections of Table 10.3 indicate that safety harness was least used in the PCWAs followed by safety shoes, highly visible clothes, and safety helmets. Interestingly, this pattern was

Table 10.2 List of factors reported to be mostly related to falls of workers from the containers at the PCWAs

Factors	Proportion of 131 workers ranked this factor as the most important one
Worn-out of the hooks and slings	55/131 (42.0%)
Not concentrated at work	41/131 (31.3%)
Oil on shoes' soles	36/131 (27.4%)
Tiredness	32/131 (24.4%)
Not following correct work procedures	29/131 (22.1%)
Disregard safety regulations	28/131 (21.4%)
Drinking beer at work	27/131 (20.6%)
Disregard safety appliances	26/131 (19.8%)
Swing slings	26/131 (19.8%)

Table 10.3 Types of safety appliance that were most frequently disregarded by the cargo handling workers at the PCWAs in Hong Kong

Safety appliances least worn by the workers	Proportion of 131 workers not using this safety appliance at work when needed
Safety harness	74/131 (56.5%)
Safety shoes	67/131 (51.5%)
Highly visible clothes	56/131 (42.8%)
Safety helmet	41/131 (31.3%)

significantly correlated with that of safety appliance least provided by the employers (Table 10.4, $r > 0.3$, $p < 0.05$, Spearman's correlation test).

The most frequently reported personal reasons for not using the safety appliances are listed in Table 10.5. Workers did not use proper safety appliances at work because they were hurrying to work (26/131 = 20%), safety appliances were not considered to be practical (22/131 = 17%), and work could be completed without using the safety appliances (13/131 = 10%). Interestingly, workers from the Stonecutters Island PCWA had a different view on this issue, they mostly considered that using safety appliances could delay their progress at work (6/30 = 20%) and also using the safety appliances at work increased the hazards of their work (5/30 = 17%). Further interviews with experts from the LCSA suggest a possible reason on how and why a safety appliance can increase the hazards. PCWA at the Stonecutters Island has the roughest sea condition among all the PCWAs and containers are constantly moving so that workers have to be very alert of their surroundings in order to avoid being hit by containers or swinging objects (e.g., slings of the cranes). The use of a helmet to reduce the severity level of head injury could block part of their field-of-views and increase the fall hazard. Also, a rigid anti-slip shoe can slow down a worker and may increase the hazard of being hit by a moving object.

Table 10.6 shows the overall results from the six PCWAs on disregarding of safety regulations. The three most frequently reported reasons for workers not following safety regulations at work were: (i) part of the safety regulations were not practical (51/131), (ii) workers could complete their work without following the safety regulations (33/131), and (iii) following the safety regulations at work could delay their progress (26/131).

Table 10.4 Types of safety appliance that were least provided by their employers (i.e., unavailable) to the cargo handling workers at the PCWAs in Hong Kong

Safety appliances least provided by the employers	Proportion of safety appliance not provided by the employers (reported by 131 employees)
Safety shoes	68/131 (51.9%)
Safety harness	57/131 (43.5%)
Highly visible clothes	38/131 (29.0%)
Safety helmet	8/131 (6.1%)

Table 10.5 Personal reasons for not using proper safety appliances when handling cargos at the PCWAs in Hong Kong

Personal reasons for not using the safety appliances	Proportion of workers that agreed on this reason
In a hurry to go to work	26/131 (19.9%)
Not suitable in real situation	22/131 (16.8%)
Work could be completed without wearing the safety appliances	13/131 (9.2%)

Table 10.6 Personal reasons of disregarding proper safety regulations when handling cargos at the PCWAs in Hong Kong

Personal reasons for disregarding the safety regulations	Proportion of workers that agreed on this reason
Part of the safety regulations were not practical	51/131 (38.9%)
Workers could complete their work without following the safety regulations	33/131 (25.2%)
Following the safety regulations at work could delay their progress	26/131 (19.8%)

Table 10.7 Proportions of near-misses happened to the workers themselves for the past year at the PCWAs. There were totally 38 near-misses reported by 131 workers

Types of near-misses	Proportion
Being hit by the containers	11/38 (28.9%)
Slip on the top of the containers	4/38 (10.5%)
Being hit by the slings	4/38 (10.5%)

10.4.7 Measurement III: Near-misses associated with the fall of workers from the top of the containers in PCWAs

Questions related to the near-misses associated with the fall accidents in the PCWAs were presented in two different ways. Workers were asked to report near-misses faced by themselves or faced by their colleagues. By asking these two questions, a more complete view on the near-misses related to the fall accidents could be obtained. As suggested by the council members of the LCSA, most of the workers at PCWAs may be over confident on their physical ability, and hence, not willing to reveal their own weakness (e.g., past records of near-misses). As a consequence, the questionnaire also asked them to report near-misses of their colleagues. The results of these two questions are listed in Tables 10.7 and 10.8, respectively. Combining the results from Tables 10.7 and 10.8, slip on the top of the containers was the most frequently reported near-misses (16/86).

10.5 Analyze phase

Results collected during the measure phase were further analyzed using the fishbone diagram and the relations diagram in accordance of the Six Sigma methodology. The fishbone diagram organizes the factors that have been reported to have caused fall accidents from the top of the containers (Figure 10.4). The clustering of various factors is based on expert opinion collected through focus group discussion with council members of the LCSA. For example, the use of worn-out hooks and slings can cause the imbalance of a container and destabilize any workers working on top of that container. When the destabilized workers have oil on shoes' soles, they can fall.

Table 10.8 Proportions of near-misses happened to their colleagues in the PCWAs for the past year. There were totally 48 near-misses reported by 131 workers

Types of near-misses	Proportion
Slip on top of the containers	12/48 (25.0%)
Being hit by slings	9/48 (18.8%)
Fall from top of the containers	7/48 (14.6%)

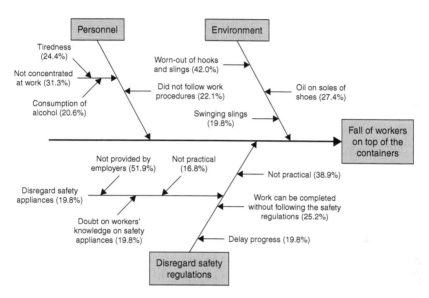

Figure 10.4 The fishbone diagram for the fall of workers from the top of the containers at the PCWAs in Hong Kong.

After clustering using the fishbone diagram, a relationship diagram is used to identify the root factors of fall accidents (HKSAR, 2003). Inspections of Figure 10.5 indicate that the following three factors have the highest number of arrows going 'out' from them: (i) not concentrating at work, (ii) disregarding safety regulations, and (iii) use of worn-out hooks and slings. These three are labeled as the suspected root factors. Among the three most hazardous work procedures identified in the survey, standing on the top of a moving container that is lifted in mid-air has the highest number of out-going arrows. Consequently, it is labeled as the most hazardous work procedure.

10.5.1 Critical factor I: Not concentrating at work

Inspections of the fishbone diagram indicate that workers could have problems on concentrating at work after working for long hours and/or drinking beer at work (Figure 10.4). This is consistent with the questionnaire data indicating

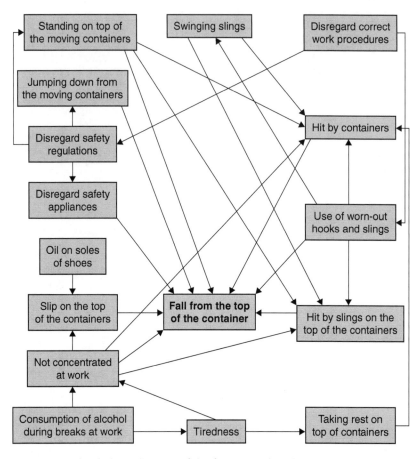

Figure 10.5 A relations diagram of the factors and work procedures related to the fall of workers from the top of the containers at the PCWAs in Hong Kong.

that 37% of the workers need to work for more than 10 h per day and 21% of the workers reported that drinking beer at work is the main cause of fall accidents.

10.5.2 Critical factor II: Disregard of safety regulations

According to Hong Kong's Shipping and Port Control (cargo handling) Regulations, Chapter 313 section B, it is the responsibility of a worker 'to wear a suitable safety helmet and use other appropriate protective clothing and equipment provided to him by an employer or person in charge of works (HKSAR, 1999). If a worker disregards this regulation and fall from a container, he may suffer a much more serious injury. In addition, it is also illegal for an employer not to provide proper safety appliances to the workers. Results of the current survey indicate that 31% of the workers do not wear safety helmet at work and 6% of their employers do not provide safety helmets. Working on top of a moving container without safety harness and safety belt was not recommended by the

Hong Kong Occupational Safety and Health Council (1999). Results of the survey indicate that 57% of workers do not wear safety harness and 44% of employers do not provide safety harness.

10.5.3 Critical factor III: Using worn-out hooks and slings

Through consultation with experts from the LCSA, experienced workers at PCWAs and a safety engineer (co-author), it has been established that the use of worn-out hooks and slings is the result of not following proper procedures. According to the proper procedure, workers are required to check the usability of the slings and hooks, and also use the correct slings and hooks for lifting containers. If workers notice that a sling or a hook is worn-out, they should report this to their employers or the person in charge so that the worn-out equipment is replaced. This cause-and-effect relationship between the disregard of correct work procedure and usage of worn-out hooks and slings are illustrated in the relationship diagram (Figure 10.5). The use of worn-out slings and hooks to lift a container can cause the container to tilt suddenly and greatly increase the chances for workers working on top of the containers to fall. This cause-and-effect relationship is shown in Figure 10.5.

10.5.4 The most critical procedure: Standing on top of a moving container which was being lifted in the mid-air by a crane

Unfortunately, standing on top of a moving container is a very common practice. The research witnessed these common practices in all visits made to all the PCWAs. Worst of all, most workers were neither wearing safety helmet nor safety harness. This practice raised great concerns in the Hong Kong Occupational and Safety Council (1999) and it would be extremely hazardous if the workers involved were not wearing safety harness to prevent fall accident.

10.6 Improve phase

According to the Six Sigma methodology, after the critical factors and hazardous procedure are identified, solutions to reduce the risks brought by these factors and procedure should be determined. The suggested solutions and root causes of these critical factors documented in this section went through thorough discussions among the research group (composes ergonomics, safety, and Six Sigma experts) and council members of the LCSA (i.e., field experts). The suggested solutions were to be implemented according to their validity. However the employers of all the cargo services companies that we have contacted refused to work on the implementation phase because they worried about the risk of exposing their unsafe practices and disrupted their existing agreements with their insurance companies. As a result, the improve scheme was passed to the Hong Kong Occupational Safety and Health Council in the hope that they, as a government agency, can persuade some companies to join the improve phase.

10.6.1 Solution I: Risk reduction for workers standing on top of the moving containers

The obvious solution is to eliminate the need for workers to stand on top of the moving containers by replacing the existing simple derrick cranes with modern quayside gantry cranes so that manual hooking-and-dehooking procedures will not be needed any more. This suggestion was rejected by the field experts because mid-stream cargo service companies could not afford the expensive gantry cranes. Also, this would bring unemployment to the stevedores at the PCWAs because this suggestion will replace mid-stream operations with modern cargo container port operations. The latter exists and is already handling 63% of the annual container throughput of Hong Kong. According to the LCSA, the mid-stream operation has its unique contributions: (i) mid-stream operations provide a very small scale and effective cargo handling service in the Pearl River delta region through the use of barges and their derrick cranes, (ii) the mid-stream operations offer a cost-effective alternative for an ocean-going vessel to download or upload a few containers without the need to dock at a modern terminal. Consequently, the following list of actions focus on measures to make mid-stream operation a safer task.

Action 1: Communications with the monitoring authority

The authority should set up guidelines to help workers to decide when it is safe for a worker to stand on top of a moving container. Currently, the practice of standing on top of the moving containers is very common even though it has caused great concern in the Occupational Safety and Health Council. It is hoped that, through this action, jobs requiring workers to be on top of the moving containers can be further classified so that some tasks are banned completely and some tasks are allowed under strict safety procedures.

Action 2: Improve the communications between the crane operator and the workers on top of the moving containers

In order to make a worker working on top of a moving container to be better prepared for any sudden container movements, it is important to improve the verbal communications between that worker and the crane operator. Currently, hand signals are used as a means of communication (Figure 10.1). This action suggests workers on top of a moving container to use hand-free communication devices to communicate with the crane operator. With the available cheap and effective wireless hand-free phones, this action is not costly.

Action 3: Easy to wear light-weighted safety harness with modified slings, and anti-slip shoes

An easy-to-use safety harness and safety belt was suggested to be worn by the workers who need to stand on top of the moving containers. The existing method was to anchor a safety belt on a stable standalone platform, however such platforms are not always available at PCWAs. The suggested method was that a safety harness could anchor to one of the four slings which were hooked to four corners of a container. The sling can be redesigned so that it has a ring or loop at a position where workers could easily hook their safety harness to it via a retractable safety belt. The authors acknowledge that this suggested action requires further testing so that the workers are not exposed to additional risks of being hit by a swing sling.

Action 4: Anti-slip shoes

Anti-slip shoes with toecap should also be worn by workers. Currently, workers are complaining that the existing anti-slip shoes are too heavy and prevent them from climbing a container. Action 4 suggests a better designed anti-slip shoe made of lighter and less rigid material that should be used for this purposes.

10.6.2 Solution II: Reduce the occurrences of workers disregarding safety regulations

The root causes of disregarding safety regulations have been identified as (i) workers consider the safety regulations as not practical, (ii) work could be completed without following the related safety regulations, and (iii) following the safety regulations could delay the work progress. In addition, the experienced workers reported during the preliminary interview that proper safety appliances were not supplied by their employers. The following actions are proposed to tackle these root causes.

Action 1: Solving problems related to confusions over the safety regulations and lack of communications among the Labour Department, Occupational Safety and Health Council and workers

We encourage the authority to arrange on-site training on the use of safety appliances for all workers. If possible, this training should be on a one-to-one basis. In addition, the authority should provide updated videotapes to demonstrate the correct usages of existing safety appliances at different PCWAs. The author, Ms. Edith Ng, attended the current safety training certification course and found that the existing course mainly focused on conceptual matters and did not provide one-to-one on-site training on the use of safety appliances.

Action 2: Toolbox safety meeting

A toolbox safety meeting is a daily short meeting between the employer and his workers to discuss the safety issues of their works. For example, the meeting can be about recent accidents or near-misses happened in their PCWA. In the case of recent accidents, the authority should send the accident report to the employers (or persons in charge) with hints on how to prevent the same accidents from reoccurring. This report should be discussed in the daily toolbox safety meetings. The toolbox safety meeting should be an open discussion between the employer (or person in charge) and the workers, and hence they could both express their opinions, concerns, and suggestions on the safety issues.

Action 3: Tackling inconvenience and delays caused by following safety regulations at work

The authors acknowledge that there is no quick fix for this problem. This action suggests the authority to set up incentive for employers and employees to get the correct balance between work efficiency and levels of safety.

Action 4: Tackling problem related to employers did not provide safety appliances to their employees

The reason why some employers did not provide the safety appliances to their employees should be investigated. If the reason is due to financial problems, the authority can set up a loan scheme for them to apply. In any case, the authority should increase inspections to reinforce the regulations.

10.6.3 Solution III: Improve the level of concentration at work

From the result of the questionnaire survey, the root causes of this problem were (i) drinking beer at work, (ii) tiredness, and (iii) prolonged working hours. Solutions for solving problems related to tiredness and prolonged working hours require another in-depth study. In the following paragraph, the improvement actions focus on the problem of beer drinking at PCWAs.

First, employers should ensure that no alcoholic drinks are stored or sold inside a PCWA. Secondly, the authority should organize a publicity campaign against alcohol consumption at PCWAs. The chairlady of the LCSA, Ms. Lam, mentioned that for the past few years the insurance fee for a normal worker working in the PCWA has already increased by 300% and now the insurance fee is around HK $3000 per worker per month. Ms. Lam hypothesizes that insurance companies will welcome a campaign to reduce alcohol consumption at PCWAs and it may even lead to a reduction of insurance premium for those mid-stream operating companies who sign up for the campaign.

10.6.4 Solution IV: Setup hooks and slings maintenance programs

According to the council members of the LCSA, there is currently no formal maintenance scheme on the lifting equipment used in mid-stream operations. In addition, there is no requirement of formal inspections of the hooks and slings used in PCWAs. Even if there is a designated examiner on hooks and slings in the PCWA, that examiner will have no formal training on how to assess a sling or hook. Consequently, the authors suggest that the authority should work with the employers to set up a maintenance system for slings and hooks. In addition, the authority should consider revising the factories and industrial undertakings regulations on suspended working platforms to cover mid-stream container handling operations.

Table 10.9 The recommended control methods for the critical factors

What is controlled?	Requirements	Control method	Frequency
Implementation of hooks and slings maintenance scheme	Clear documentation of the routine maintenance scheme Designate trained persons in charge of the maintenance scheme	Employers to ensure the scheme was implemented Inspection from the authority	Daily inspection by the employers is suggested Monthly inspection by the authority is suggested
No alcohol consumption at work	No alcohols are allowed to be stored inside PCWAs No alcohol should be sold in the canteens inside PCWAs	Inspection by the employer and authority	Daily inspection by the employers is suggested Monthly inspection by the authority is suggested

10.7 Control phase

To prevent a fall accident and its critical factors from reoccurring, control methods are needed. Suggested solutions included engineering solutions on fall protection (e.g., new designs of safety harness and shoes), administrative solutions on training (e.g., toolbox safety meeting), and suggestions to improve cooperation among authorities, employers, and workers. As explained above, no company is willing to participate in the improvement phase, as such, the following control methods are only suggested plans.

In Table 10.9, two control plans are suggested in order to sustain the improvements on hooks and slings maintenances and no alcohol consumption at work. For sustaining the improvements, both the authorities and employers are required to conduct inspections. Employers, or their designated persons, are suggested to conduct daily inspections. In addition, the results of inspections should be documented. Any problems found during inspections should be discussed during the daily toolbox safety meetings. The authority should inspect the progress of the improvement plans on a monthly basis. In addition, the authority could also discuss with the employers on how to sustain the improvements which could also help to improve the connections between the authority and the industry.

Through these improvement schemes, chances of using worn-out hooks and slings are expected to reduce rapidly. In addition, occurrences of workers drinking alcoholic drinks should also be reduced. The expected consequence is a greatly reduced rate of fall accidents at PCWAs.

10.8 Conclusion

The critical factors and the most hazardous procedure related to fall accidents at the PCWAs in Hong Kong have been identified. From the quantitative procedures outlined in the Six Sigma methodology, we identified the critical factors from the questionnaire survey as follows: (i) lack of concentration at work,

(ii) disregard of safety regulations, and (iii) use of worn-out hooks and slings. The most hazardous procedure that can lead to fall accidents from the top of the containers is 'workers standing on top of the moving containers that are being lifted by cranes.' The survey has covered all the PCWAs in Hong Kong that handle containers. The root causes of these critical factors and hazardous procedures have also been determined through focus group discussion with the council members of the LCSA and safety engineers. Solutions to reduce the causes are documented in the chapter. Examples of engineering solutions include customized fall protection systems and light-weight anti-slip shoes. Examples of administrative solutions include daily toolbox safety meetings, measures to prevent alcohol consumption at work, and a formal equipment maintenance scheme. Measures to facilitate better cooperation among the authorities, employers, and workers have also been suggested to clarify safety regulations and the use of safety appliances. In addition, the authorities have been encouraged to reinforce the legal requirement of employers providing safety appliances to their employees. These solutions have been documented and passed to the LCSA and the Hong Kong Occupational Safety and Health Council.

Currently, fall accident problems are handled through qualitative approaches that rely on experience and gut feelings without much quantitative measurement and analyses. The reported application of DMAIC provides a systematic and data-driven alternative approach to tackle these problems (Pande *et al.*, 2002). The case study demonstrates the effectiveness of the Six Sigma DMAIC approach to reduce fall hazards and also indicates that such a rigorous approach can be a means to meet increasingly high safety standards and an eventual near-zero hazard rate.

Acknowledgements

The authors are grateful to the editor and the anonymous referees for their valuable comments. The authors would like to thank the Hong Kong Logistics Cargo Supervisors Association (LCSA) for support and help in this study. The work was supported by the Hong Kong Occupational Safety and Health Council through grant HKOSHC 01/02. EG01.

References

Hong Kong Port and Maritime Board (2002). *Hong Kong Port and Maritime Board Annual Report 2002.* PRC: Hong Kong Port and Maritime Board, Hong Kong Special Administrative Region (HKSAR).

Hong Kong Special Administrative Region (HKSAR) (2002). *Invitation of Research Proposals in Strategic Areas of Importance in Occupational Safety and Health in Hong Kong.* Hong Kong: PRC: Hong Kong Special Administrative Region (HKSAR), Hong Kong.

Hong Kong Special Administrative Region (HKSAR) Marine Department (2003). New cargo area in Chai Wan to take over Wan Chai operations, 7 July. http://www.mardep.gov.hk/en/publication/pressrel/pr030707.html

Hong Kong Special Administrative Region (HKSAR) Marine Department (2003). Casualties in cargo handling accidents (1992–2002), 16 September. http://www.mardep.gov.hk/en/publication/mias.html#a1

Hong Kong Special Administrative Region (HKSAR) Marine Department (2003). Port services, 23 October. http://www.mardep.gov.hk/en/pub_services/ocean/pcwa.html

Hong Kong Special Administrative Region (HKSAR) Occupational Safety and Health Council (1999). *Transport and Cargo Handling GuideBook.* PRC: Hong Kong Special Administrative Region (HKSAR). HKSAR Occupational Safety Council, Hong Kong.

Ng, E., Tsung, F., So, R.H.Y., Li, T.S., Lam, K.Y. and Tang, C.T. (2003). Hazards to workers working on top of cargo containers: critical operations and factors. *Proceedings of Safety and Health Expo*, 18–20 March, Hong Kong.

Litwin, M.S. (1995). *How to Measure Survey Reliability and Validity.* CA, USA: Sage Publications.

Pande, P.S., Neuman, R.P. and Cavanagh R.R. (2002). *The Six Sigma Way Team Fieldbook.* NY, USA: McGraw-Hill.

11

Six Sigma for Government IT: strategy and tactics for Washington, DC

Rick Edgeman, David Bigio and Thomas Ferleman

11.1 Introduction

Information technology (IT) has altered the way businesses and governments interact and make decisions and many municipal governments enable their citizenry via IT. While IT is expected to add efficiency, increased reliance on IT demands highly reliable service with assurance that online information is secure, accurate, and available almost instantly so that server problems must be prevented or, on occurrence, rapidly detected and corrected. Municipal governments should also be good stewards so that financial management (FM) is critical. These issues are perhaps nowhere more evident than Washington, DC, where the Office of the Chief Technology Officer (OCTO) was forged in direct response to years of IT disinvestments and charged with consolidating the fragmented IT systems of DC's 68 agencies into a standardized infrastructure and solving the District's IT system budgeting and planning problems, to transform DC into the *City of Access*.

OCTO's human capital is generally unfamiliar with and untrained in use of quality tools or strategies. Moreover, personnel is already *stretched*, hence OCTO engaged quality enhancement systems and teams (QUEST) consulting teams from the University of Maryland to conduct a comprehensive analysis of IT service level management (IT SLM). Five IT SLM areas were examined through two lenses with a team assigned to each area and lens. The IT SLM areas examined were availability management (AM), capacity management (CM), FM, service continuity (SC), and SLM. The two lenses were *Six Sigma* and *Business Excellence*. QUEST teams derived a number of strategic and tactical improvement suggestions which aimed at advancing OCTO in its endeavor to aid DC in becoming the *City of Access*. Presented herein are highlights of the effort with results enhanced by synergy between the two lenses. OCTO leadership projects between \$2 million and \$3 million in long-term savings from these efforts.

11.2 Motivations

Imagine a well-timed terrorist strike crippling or compromising the integrity of the District's IT infrastructure. Such possibilities are thinkable in light of the 11 September 2001 strike on the Pentagon, immediately across the Potomac River from Washington, DC only a few kilometers from the White House and OCTO headquarters. Secondly, public sector entities be wise stewards publicly supplied resources that enable their efforts. One reaction to this has been the 'slash and burn' of eliminating activities and people. Alternatively an organization can become dramatically better by better understanding customer needs and improving processes to fulfill them, an approach ideally suited to both *Six Sigma* and *Business Excellence*, and their integration with *Lean Management* practices. Thirdly, while OCTO currently provides IT resources to District agencies free of charge, future reality may allow for fee-based service provision. What is tolerated 'for free' and what is demanded in exchange 'for resources' often differ. Soon, agencies currently resourced by OCTO may freely negotiate with competing providers while OCTO may be required to provide only minimal services, hence a client hierarchy based on more than the criticality of agency functions may emerge wherein OCTO must reliably and cost effectively provide IT services for which client agencies may find themselves competing.

In this light QUEST examined IT SLM as driven by the Information Technology Infrastructure Library (ITIL) best practices suite. ITIL encompasses SLM, FM, CM, AM, and SC where for practical purposes SC often manifests as 'disaster recovery.' Enterprises worldwide have implemented ITIL to improve delivery of IT services with Microsoft, Price Waterhouse Coopers, and Hewlett-Packard among users. Teams first focused on 'ITIL maturity' assessment of each ITIL area in a manner akin to the self-assessment notions of *Business Excellence*. These assessments informed the initial phases of *Six Sigma's Define–Measure–Analyze–Improve–Control* (DMAIC) or *Design for Six Sigma* (DFSS) approaches. QUEST emphasizes team and organizational effectiveness along with strategies and methods that enhance product, process and, system quality. QUEST teams function in a project-rich environment and complete projects for organizations of all sizes and business sectors. OCTO sought to leverage QUESTs structure, expertise, and experience generally and, especially as relates to *Six Sigma*. A council of OCTO and QUEST representatives was charged with integration and subsequent deployment of recommended improvements. Results for SC and AM are presented in some detail herein. Remaining areas discussed only briefly.

11.3 Six Sigma and quality concepts

Many *Six Sigma* methods are statistical ones long in use or management and planning ones in vogue wherever teamwork and breakthrough thinking are valued. While *Six Sigma* is often highly statistical, it is also strategic in its orientation and project nuances dictated that some phases were conducted in this strategic, more general, less statistical sense. OCTO was ripe for application of *Six Sigma* since it supports agencies that serve critical societal or environmental functions. Consistent with the *Pareto Principle*, it was noted early on that a few agencies dominate, filling especially critical functions, something of which OCTO was only anecdotally aware.

Six Sigma and *Business Excellence* emphasize exceptional performance or '*excellence*,' an overall way of working that balances stakeholder interests and increases the likelihood of sustainable competitive advantage through operational, customer-related, marketplace, and financial performance excellence (Edgeman *et al.*, 1999). In part this requires (self) assessment – the regular, rigorous, and systematic review of an organizations' approaches, deployment, and results thereof in order to determine strengths, weaknesses, and areas where resources should be applied. While assessment reveals key weaknesses, it is non-prescriptive – leaving remedies to the wiles of the organization. *Six Sigma* endeavors identify and deploy best solutions to important issues, attaining outstanding results. Generally, ITIL assessments identified key issues whereas *Six Sigma* generated solutions so that *Six Sigma* served to enable *Business Excellence* and assisted in more clearly resolving and fulfilling the voice of client agencies – the so-called '*Voice of the Customer*' (VOC). Project efforts enabled OCTO to improve internal performance and enabled agencies to better serve the citizenry. Klefsjö *et al.* (2001) discuss relationships between *Six Sigma* and initiatives such as *Business Excellence*. Extensions and adaptations of *Six Sigma* are proffered in Edgeman and Bigio (2004).

Six Sigma organizations apply DMAIC or DFSS structured knowledge-acquisition/problem-solving strategies that develop superior ideas leading to superior results in areas of strategic import, including financial results. DMAIC and DFSS are data-driven, fact-based approaches emphasizing discernment and implementation of the VOC. *Six Sigma* employs highly effective methods of VOC discernment in the *define* phase of both DMAIC and the DMADV (Define–Measure–Analyze–Design–Verify) algorithm used in DFSS. Various tools and techniques are used throughout *Six Sigma* projects with those used in the OCTO effort such as quality function deployment (QFD), the house of quality (HOQ), and failure modes and effects analysis (FMEA) summarized in Table 11.1.

11.4 The ITIL

ITIL focuses on providing high-quality services, emphasizing customer relationships to aid OCTO in fulfilling provisions of its Service Level Agreements (SLAs) with client agencies. SLAs outline service charges and contractually bind an agency and OCTO. Service delivery concerns formation of SLAs and compliance-based performance monitoring. On the operational level, service support processes address changes needed and any service provision failures. ITIL provides defined processes and best practices for IT services

Table 11.1 Summary of Six Sigma tools and approaches used to enhance perform-ance at OCTO

Approach	Description or example use	ITIL areas
Charter	Purpose is to define the business case, project goals and limits, way of working together, and conflict resolution plan.	All
Brainstorming	Uses included cause identification and solution generation.	All
Affinity diagram	Uses included associations among OCTO needs (critical to qualities (CTQs)).	All
Inter-relationship digraph	Primarily used to explore causal relationships between enablers ('hows') to capture correlations and form the roof in the HOQ.	AM, CtM
Nominal group technique (NGT)	NGT used as part of QFD/HOQ to prioritize OCTO needs.	All
Matrix/priority matrix diagrams	Various uses including distribution of tasks to team members and relating OCTO needs (CTQs) to enablers ('hows') in QFD.	All
SMART goals	*S*pecific, *m*easurable, *a*ttainable, *r*elevant, and *t*ime-bound goals and problem statements.	All
Process maps	Included high-level COPIS maps and detailed process maps. Both *before* and *after* improvement versions were used.	All
VOC tools	Approaches included surveys, focus groups, customer complaints, and interviews.	All
Drill down trees	Process-product drill down tree.	All
FMEA		AM, CtM
HOQ/QFD	Integrated use of matrix diagrams and NGT to assess internal and external customer needs and deploy solutions	AM
Pareto chart	Used to identify dominant issues/defect causes.	All
Fishbone diagram	Also called cause-and-effect diagrams display 'effects' representing a problem or an opportunity with the 'causes' being real or potential drivers of the effect.	All
SWOT	All ITIL areas were assessed for *s*trengths, *w*eaknesses, *o*pportunities, and *t*hreats with motivations being improvement or leverage of strengths, diminution of weaknesses, welcoming opportunities, and countermanding threats.	All
Benchmarking		CtM
Chi-square	Tests of homogeneity examined whether differing approaches yielded similar results and tests of independence explored CTQ–enabler relationships.	AM, CaM, CtM, SLM
Correlation and regression	Used to explore, assess, characterize, and exploit CTQ–enabler relationships.	SLM
DOE	Design of experiment approaches included experiments with operating parameters, critical elements, or both. Operating parameters are enablers (X's) that vary in amount while critical elements are enablers (X's) that differ in type or categorically. Screening and factorial designs were used.	AM, CtM, SLM
SPC charts	Statistical process control charts recommended or used included p charts, I–MR charts, and X-bar and R charts to (directly) control the X's, hence indirectly the CTQs.	AM, CaM, CtM, SLM

(continued)

Table 11.1 (Continued)

Approach	Description or example use	ITIL areas
Savings	Total savings estimated by Deputy Director = $2–$3 million from 2003 to 2007	
ITIL areas	A: availability management; CaM: capacity management; CtM: continuity management; FM: financial management; and SLM: service level management	

management – a fast track toward documentation levels and certified perform-ance akin to those of ISO 9000. ITIL benefits include:

- Improved service quality.
- Cost justifiable service quality.
- Services that meet business, client, and user demands.
- Integrated centralized processes.
- Everyone knows their service provision role and responsibilities.
- Learning from previous experience.
- Demonstrable performance indicators.

ITIL is public domain, may be adopted in proprietary ways, and is widely regarded as the only consistent and comprehensive documentation of IT service management best practices. Each ITIL module promotes management of an organization's IT services and infrastructure where 'IT infrastructure' means computers and networks, hardware, software, and computer-related telecom-munications on which systems and IT services are built and run. ITIL codes of practice assist an organization in fulfilling current and future IT service require-ments in the face of budgetary constraints, skill shortages, system complexity, and rapid change.

SLM ensures that SLAs are met and that adverse impacts on service quality are minimized, assessing the impact of changes on service quality and SLAs, both when changes are proposed and after their implementation. Key targets set in SLAs relate to service availability thus requiring incident resolution within agreed periods. SLM is the hinge of service support and delivery and relies on the effective and efficient working of underpinning support processes, without which an SLA is useless, since these are foundational to content agreement.

CM ensures constant availability of adequate capacity to meet agency busi-ness requirements. CM involves incident resolution and problem identification for those difficulties related to capacity issues and generates requests for change (RFC) that ensure sufficient capacity. RFC are subject to a change man-agement process and implementation often affects hardware, software and documentation, and requires effective release management.

AM concerns design, implementation, measurement, and management of IT services to ensure that stated availability requirements are met and requires IT service FMEA and the understanding of the time taken to resume service. Incident management and problem management provide key inputs ensuring that appropriate corrective actions occur. Availability targets specified in SLAs are monitored as part of the AM process that also supports the SLM process by providing measurements and reporting to support service reviews.

IT service continuity management (SCM) or *disaster recovery* manages an organization's ability to provide a pre-determined agreed on level of IT services to support *minimum* business requirements. Among the means used are resilient systems and recovery options such as backup facilities. Configuration management data is required to facilitate this prevention and planning. Infrastructure and business changes need to be assessed for their potential impact on continuity plans, and IT and business plans are then subject to change management procedures.

FM accounts for costs and returns of IT service investments and cost recovery from clients. FM requires interfaces with CM, configuration management, and SLM to identify the true costs of service. FM works together with business relationship management and the IT organization during the negotiation of IT budgets and client IT expenditures.

In ITIL, SCM involves developing backup systems, recovery processes, redundancy, and contingency plans to provide agreed on levels of IT services to support clients' minimum business requirements. OCTO actions effect agencies and residents, thus making IT SC vital. DC's IT infrastructure benefits citizens by increasing the accuracy of their records and access to government services. Integrated databases, centralized processing, and web-based user interfaces reduce conflicting records among agencies and allow citizens to interact with agencies without leaving their home or office. Without sufficient backup and recovery mechanisms, it is possible for corrupted files, terrorist attacks, or other massive malfunctions in data processing and storage to impair citizen interactions with the system. Continuity is important to the agencies since they do not have their own processing centers and rely on centralized processing and a high-speed wide area network (WAN) for most transactions. The IT infrastructure in DC is configured so that all agencies and their data and communications are transmitted along a single WAN, all data processed by a single pair of data centers, and all data stored in a single pair of data storage centers. DC's traffic lights and emergency call centers are connected to the WAN as is the wireless network for police, fire, snow removal, and garbage trucks. The WAN is the backbone of the city.

11.5 ITIL maturity

The primary assessment goal was to identify areas of the ITIL service delivery framework most in need of attention to advance ITIL maturity. Assessments examined that the nine progressive elements of IT process management cited an ITIL framework which intended to provide high-quality IT services. To assess OCTO's position relative to the framework, maturity survey questions addressing the nine progressive elements of each ITIL area were answered:

- *Prerequisites* determine whether a minimal array of prerequisite items is available to support process activities.

- *Management intent* establishes whether there are organizational policy statements, business objectives (or similar evidence of intent) providing both purpose and guidance in the transformation or use of the prerequisite items.
- *Process capability* examines activities being carried out. Questions are aimed at identifying whether a minimum set of activities are being performed.
- *Internal integration* seeks to ascertain whether activities are integrated sufficiently to fulfill the process intent.
- *Products* examine actual process output to enquire whether all relevant products are being produced.
- *Quality control* concerns review and verification of process output to ensure that it is in keeping with the quality intent.
- *Management information* concerns governance of the process to ensure provision of adequate and timely information sufficient to support necessary management decisions.
- *External integration* examines whether all external interfaces and relationships between the discrete process and other processes are established within the organization.
- *Customer interface* is concerned with on-going external review and validation of a process to ensure that it remains optimized toward meeting client needs.

Failing results occurred in *all* ITIL areas in each case with respect to a majority of the nine elements. It was in some sense not so desperate as it might be assumed in that the human resource at OCTO is both exceptional and motivated. Generally, assessments pointed out deficiencies related to measurement, and the creation and documentation of formal processes whereas remedies improved the nature, quality, measure, and communication of information between OCTO and client agencies or connectivity of ITIL areas.

11.6 Assessment and improvement of SLM

SLM performance monitoring is the process by which OCTO measures, analyzes, and reports on how well they are meeting objectives outlined within SLAs. The motivation for improving the performance monitoring process comes from the discrepancy between OCTO's perspective of service performance and their customer's perspective of services received. Real and potential benefits of SLM include enhanced service quality, customer relationships, and financial position. Analysis revealed that OCTO personnel did not use a standard performance monitoring process and lacked sufficient understanding of ITIL SLM best practices or compliance therewith. Poor performance relative to ITIL criteria was not a surprising finding since there was little understanding of the specific information and flow thereof needed to inform OCTO management at a level enabling meaningful service improvement and communication with client agencies. Lacking a standard SLA with defined content and a means of assessing conformance thereto, there was no single 'real' process map since SLAs were constructed in a 'one off' manner. To remedy this situation the recommendation was made to OCTO to structure or identify needed content and form for a standard SLA, whereon OCTO identified its SLA with the Office of Tax and Revenue (OTR) as the future conformance standard. This SLA was analyzed with the following nine categories of information therein identified:

- *Availability*: how the SLA defines customer accessibility to the network.
- *Sub-system performance*: how OCTO evaluates operational performance.
- *Host response time (HRT)*: host time spent on processing.
- *Batch services*: central processing unit (CPU) limit, turnaround, and response time for batch jobs.
- *Network services*: responsibilities of OCTO's Network Operations Center.
- *Helpdesk support*: responsibilities of OCTO's helpdesk.
- *Technical and operational support*: OCTO's responsibilities to provide customers' technical and operational support.
- *Disaster recovery*: OCTO's role in the disaster recovery plan (DRP).
- *Service level reporting (SLR)*: OCTO's responsibilities in reporting service levels to customers.

Using the OTR SLA as the standard for future SLAs, relevancy of data collected relative to the standard is key OCTO's performance monitoring process so that ensuring collected data is relevant to performance is critical. Analysis identified availability, performance, and reliability as key SLA metrics with all collected data correlating to these. OCTO collects and analyzes data from 24 categories with 14 correlating to availability, 4 to performance, and the remainder to a mix of performance and reliability. As one project outcome a means of assessing new data types relative to the three SLA metrics was created, evaluating each new data type against a checklist determining how data fits into the scheme and filtering out unneeded data. A key project deliverable resulting from this effort was an 'ideal process' map for SLA creation and monitoring of compliance to the provisions therein. OCTO implementation of the corresponding ideal process standardized, streamlined, and improved the performance monitoring and reporting process thereby improving service delivered to agencies and agency satisfaction.

11.7 Assessment and improvement of CM

IT infrastructure is costly to upgrade. Enterprise software is licensed according to the power of the machine on which it runs – the principle being that more power facilitates more powerful use so that upgrading a mainframe computer has both hardware costs and larger software licenses costs. CM seeks to maximize the power of the existing configuration and ensure that the IT resources supply is sufficient to satisfy client demands. This requires accurate assessment of current and future client needs to negotiate SLAs. Critical to quality (CTQs) characteristics identified include ITIL implementation, development of a structured capacity plan, assurance of compliant SLAs, accurate forecasting of IT trends and impact thereof on service needs, and development of accurate process maps linking OCTO capabilities to agency work flow. Implementing ITIL promoted effective use of ITIL infrastructure while compliant SLAs ensure that capacity meets or exceeds agency needs. This demands clear, timely communication of information with OCTO failing to attain needed levels.

Assessment of the VOC and use of QFD, cause-and-effect diagrams, and process maps clearly pointed to the importance of and deficiency in communication within OCTO and between OCTO and client agencies. Subsequently a communications plan was developed and deployed that enhanced understanding of the balance between resources and the activities to which they are allocated. Early results of these efforts indicated significant improvements in the management information and external integration elements of the ITIL maturity, and hence both internal and external improvements.

11.8 Assessment and improvement of FM

While detailed discussion concerning ITIL FM and its assessment could be provided, a general preliminary assessment revealed a severe lack of standardized financial and accounting practices with insufficient granularity so that budgeting at OCTO involves significant 'guesswork.' Indeed, OCTO decision-making processes requiring accounting information are nearly non-existent so that detail needed to support either activity-based costing (ABC) or creation of a balanced scorecard does not exist. Additionally, there is a lack of a common key across accounting spreadsheets that would allow, for example, sorting of data by invoice number or transaction number; there are multiple names assigned to the same product; and costs are not always assigned to the activity that incurred the costs. Charging for IT services implies charging agencies for the services rendered to them. OCTO is not at the maturity level necessary to charge for any services. As OCTO is in many ways an 'in-house' organization for various DC agencies, charging for services remains a lesser priority at this juncture. That is fortunate since the cited problems are a reflective – not exhaustive – listing of the state of OCTO's FM systems and processes. Creating sound infrastructure is the main concern at this point so that work focused primarily on creation of a common naming convention, ABC, and structuring of key processes along with documentation thereof including process maps. It is accurate to characterize the approach employed as one making use of many tools common to *Six Sigma* projects such as the VOC, process maps, cause-and-effect diagrams, and QFD. Through use of these tools the general result of the work performed is that vital connections of FM to the rest of OCTO and especially the remaining ITIL areas were made more explicit and a number of uniform accounting, budgeting, charging, and reporting processes were developed and documented. A natural consequence of OCTO following through on the work performed is that significant progress toward ITIL FM maturity can be expected. The work performed clearly indicated that OCTO can realize dramatic improvement in its FM IT infrastructure by addressing the ITIL assessment areas of process capability, internal integration, quality control, and management information.

11.9 Assessment and improvement of AM

AM is concerned with the design, implementation, measurement, and management of IT services to ensure that stated business requirements for availability are consistently met. This requires understanding of FMEA and the time taken to re-establish service after a failure. The overarching project goal was to improve IT infrastructure efficiency by limiting the frequency and duration of IT service outages, and hence increasing availability. AM addresses three key areas: overall system availability, system downtime, and response time. OCTO has limited control over the networks maintained by the various agencies they serve so that it is vital that needed information from the OCTO mainframe must be available. Data must be collected determining the percent availability of the system. As regards system downtime, it is important to be able to answer such questions as: 'where and when did the downtime occur?' Correctly answering these questions provides data that may identify key system deficiencies that can be corrected to provide maximum availability. Addressed response time issues included 'how quickly the system becomes available once reported down and response time reduction.' Improved reporting schemes can improve efficiency and provide the data necessary to understand a problem, since once understood it can be better resolved. Optimizing these aspects improves system performance. The AM project included maturity level assessment and development of a strategy better aligning OCTO with ITIL best practices.

11.9.1 AM define and measure phases

OCTO representatives and their client agencies were the primary and secondary 'customers,' respectively, through which the VOC was discerned through regular and rigorous dialog, surveys, and correspondence. Processing the VOC led to identification of several CTQs that were weighted by importance and provided inputs to more formal evaluation via the HOQ. A partial list of key actions and direction of improvement reflecting the VOC included decrease system downtime; increase percent availability of system; conformance to SLA with agencies; decrease frequency of system downtime; increase system speed; decrease problem response times; increase system performance consistency; automate system problem detection; increase scheduled system downtime information and its flow; establish troubleshooting procedures; increase unexpected downtime information; and increase availability performance reporting. Also determined were quantifiable criteria and metrics. These include amount of scheduled and unscheduled downtime and frequency of each; time required to detect failures and identify failure mode; problem response time; data transmission speed; transmission speed standard deviation; system failure time; IT staff size; and proximity of IT staff to the client. Identified by the HOQ as most critical were unexpected downtime, amount of scheduled downtime, frequency of downtime, and failure detection and resolution time. This information focused on subsequent project efforts.

Measurement focused on distributing, compiling, and honing an ITIL AM practice survey. A standard survey can be obtained from the ITIL website at: http://www.itil.co.uk/. Completion of this survey purports to reveal AM maturity levels in the nine progressive areas defined previously. Similar maturity assessment surveys for the remaining ITIL dimensions are also posted at the website. While survey questions only permit *yes* and *no* responses, included are questions to which finer gradations are both possible and valuable as well as compound questions that still allow only a single *yes* or *no* response. These and other shortcomings in the survey motivated development of a revised survey instrument to address these inadequacies and, subsequently, to provide fuller maturity information. While performing most effectively with respect to process capability and external integration, even these areas received failing marks.

11.9.2 AM analyze phase and benchmarking

Assessment results were analyzed and improvements that would bring OCTO closer to ITIL AM maturity in weakest areas were made. Analysis indicated that significant progress could be made by broadly informing personnel of the meaning, purpose, and benefits of availability; creating an availability plan (scope, procedures for monitoring, analyzing, and forecasting availability); assigning responsibilities for AM activities; training relevant personnel; compiling availability reports with performance evaluation and providing them to relevant areas of the organization; enhancing customer relationships; and monitoring customer satisfaction trends.

The services provide by OCTO are by no means unique in that various IT shops provide similar services to their clientele and hence OCTO could benefit from benchmarking. Benchmarking is used by organizations globally to measure, improve, and redefine their processes. OCTO provides DC agencies free Internet and other service. Since OCTO expects to begin charging for their services, agency expectations of and relationships with OCTO are certain to change with one possibility resembling a private company paying for service from an Internet service provider (ISP). Increasing dependence on technology-driven business solutions fuels the need for stable and reliable backbones, a result of which has been proliferation of ISPs – some of which will have processes that, adapted to OCTO's environment, could fuel needed improvement. These agencies conduct business every day assuming that needed data will be instantly available. Strategic benchmarking may catalyze a paradigm shift at OCTO, the potential benefit of which is imitation of the best practices and core competencies of the world's leading ISPs and be responsible to its customers, providing quality service at competitive prices.

11.9.3 AM conclusions

OCTO must take important steps in five areas: personnel management, training, planning, process monitoring, and communication. Completing all recommended tasks will advance ITIL AM maturity. As regards, *personnel management* OCTO must assign AM activities to specific individuals; create a single point of accountability or process owner for CM; and ensure that other specific ITIL groups are created. In terms of *training*, OCTO should detail and disseminate information describing the importance of AM and its standards and criteria; and develop a uniform training program for relevant employees. To satisfy its *planning* goals and objectives OCTO must identify and document availability requirements; define the scope of AM; establish preventative measures; set goals that can be quantitatively and qualitatively assessed and regularly reviewed; and create an availability plan. A significant portion of ITIL AM maturity depends on *process monitoring* and reporting. To advance ITIL maturity OCTO should develop procedures for monitoring, analyzing, and forecasting service availability; record data on service availability and component failure; record data on response times; make recommendations for improvements in IT service availability; issue standard reports on IT service availability; and assess changes as they are implemented with further monitoring. Further, ITIL AM maturity is deficient in part due to poor *communication* which *must* be enhanced across ITIL areas.

11.10 Assessment and improvement of SC

Preparedness for power outages, excessive resource demand, and the like – even terrorist strikes – is necessary. Apart from DC's political, economic, and military importance, it is second only to Los Angeles among traffic-congested US cities, so that one can imagine the effect on the city if all traffic signals were rendered inoperable and, while such an event might be related to the power grid, this could result from IT failure. One can posit other scenarios related to the police force, or public education, or fire and rescue and so on: IT SC is critical and rapid recovery from any interruptions of such service is a must under many scenarios, both foreseen and unforeseen. More generally, SC is concerned with an organization's ability to provide a specified level of IT services to support its minimum business requirements. This implies that if something fails there is a backup plan in place that will allow an organization's IT infrastructure to function so that business processes can continue. There are three main facets of IT SC: backup and redundancy, risk management, and change management. These parts of SC ensure that when disasters occur, IT systems will still function or that service will be quickly recovered.

Backup and redundancy are technical aspects of SC. They are tangible parts of an IT network that you can physically see and it is important that they are ready to perform. Backup systems have a number of meanings, sometimes overlapping with redundancy. Backup systems are fully functional systems ready to immediately takeover for a failed primary system. OCTO has created a system where there are two different data centers, each with its own mainframe – with mirrored disk drives so that if one of the mainframes fails, all needed data can quickly and seamlessly be accessed from the surviving mainframe. Backups are also methods of archiving and storing data from a server onto media and allow data to be stored on several volumes of media that can quickly be accessed and used to restore data that has been lost or corrupted. Similar to backups, redundancy involves having extra layers of IT to fall back onto if primary systems fail. Redundancy implies ability to complete the same task in multiple ways simultaneously so that data centers can communicate through the redundant/backup WAN links. In addition to having multiple WAN links most IT data centers have multiple redundant power supply systems. While data centers receive their power from the power company they also have uninterruptible power supply (UPS) systems on site. These are, in essence, large batteries or generators that can power a data center for extended periods during a power outage. Redundancy also refers to multiple layers of physical security that discourage information tampering or unauthorized data center entry.

While backup and recovery methods provide insurance when disasters have occurred, risk management (RM) analyzes an IT department for vulnerabilities and derives plans to minimize these and prevent their occurrence. RM outlines how to respond to the exploitation of any known vulnerabilities and deals with both technical and non-technical issues. Technical issues include security vulnerabilities related to the operating system, or software running on a data center's computers that can be exploited by unauthorized individuals. RM also looks to eliminate single points of failure (SPOF). If all data is run through one computer and then routed to where it needs to be stored, that one computer is a potential SPOF. Risk analysis (RA) is a systematic method of identifying the assets of a data processing system, the threats to those assets, and the vulnerability of the system to those threats. Information gleaned from RA allows an organization to determine risk mitigation measures and can be helpful in developing and organizing crisis response teams. After vulnerabilities are identified, management must make sure that systems are improved to lessen the chance of exploitation. Change management involves thoroughly analyzing any and all changes that are made to IT systems to ensure that any impact of these changes can be mitigated quickly. Specific procedures should detail any changes and an 'escape route' that can undo poorly implemented changes or ones causing unforeseen problems should be mapped out.

11.10.1 Define, measure, and analyze phases for SC

ITIL SC maturity was assessed and found wanting so that processes supporting SC and advancing ITIL maturity were designed. The work was limited by a paucity of data and the fact that no uniform disaster recovery process existed. As such the approach taken was more or less a hybrid of DMAIC and DMADV, and almost immediately delivered low hanging fruit in the form of a process map, followed by derivation of a data collection methodology that would examine customers, OCTO, data collection, SLAs, and internal problems. This required 'defect definition' and, in consort with OCTO expertise, a 'defect' was defined as the capacity for inefficiency within the SC plan. Defects occur when continuity structures are not allocated based on conditions of an SLA and systematic elimination of defects was targeted since, absent defects, an SLA can be regarded as met. Maturity assessment indicated deficiencies in all areas but especially with regard to prerequisites, management intent, process capability, and internal integration. CTQs were identified with the notion that focusing on these would drive OCTO toward ITIL SCM maturity. The CTQs developed were definition of minimal SCM functions; definition of business critical procedures; establishment of a service recovery procedure; and establishment of a testing procedure. Minimal SCM operational functions were defined to create basic requirements of OCTO's future SCM plan. These included minimum network capacity, workforce needed to maintain daily operation, occurrence rate, and severity of service interruptions. An SC process map was developed that enabled a more efficient and effective business procedure that minimized user confusion during service interruptions.

Six Sigma is data driven and the maturity assessment provided much of the data necessary to develop an SC recovery process map as well as key improvement recommendations and metrics for areas lacking with respect to ITIL SCM maturity. The process map provided a foundation for SC growth. Each metric requires action to attain acceptable performance levels and the HOQ was used to identify correlations between metrics and requirements. In the prerequisites area it was discovered that SCM leadership needed to be established and that minimum operational requirements needed to be determined. In the management intent area it was determined that business impact need to be analyzed, that the SCM procedure was in need of testing and that resources for SC plans needed to be provided. Last, in the process capability area, it was determined that impact analysis should be used to determine minimum business critical requirements, a coordinated implementation plan should be established, assessment should be conducted, IT SCM components for business continuity need to be identified, a checklist for covering actions required for all stages of disaster recovery should be established, and testing procedures should be established. Existing procedures for managing problems were confusing and development of a simplified process enhancing the functionality of the helpdesk and assuring meaningful collection of user issues were needed. The developed process enabled consistent data acquisition, analysis, and resolution to disruptions in IT service.

11.10.2 Continuity design phase

Internal examination provided limited useful information. This led to examination of other strategies used to attack the same goal. Toward that end the DRP and procedure of the South Dakota Bureau of Information Technology (SD BIT) was benchmarked for adaptation of their SCM plan. The SD BIT plan included sections on customer service and disaster recovery. The SD BIT works with the customer to drive the problem tracking system and recovery procedures. This pointed to the need for a more effective OCTO–client interface. The overall SCM process can be improved by using data to match the user's perception to measured performance values. Currently there is little established data thus limiting the capabilities of the *Six Sigma* approach. Self-assessment can be used to continuously benchmark internally. Resource savings can be tracked by comparing questions that pass with questions that fail. Such tracking may aid the determination of the relative importance of user needs. Comparing user satisfaction between agencies is feasible if each agency conducts its own assessment. Initially such data will reveal strengths and weaknesses of current practices between OCTO and each agency. SLAs can be added to the factors used and weighted into evaluations. Since SLAs dictate what an agency should receive, awareness on the part of an agency and OCTO is important in order to improve SCM. Finally, by tracking the problems received by the helpdesk and comparing these to SLAs and the self-assessment, a multiple factor data collection becomes feasible thus allowing SCM to more rapidly approach higher levels of service. Optimization of these three factors will require time and other resources but will provide better long-term results.

11.10.3 Continuity verify phase

Since the *verify* stage requires prior implementation by OCTO, only recommendations were possible, the first of which was development of an implementation strategy for SCM growth. SLAs were reviewed as were the maturity assessment and preliminary draft of the continuity plan. From these the following were made. The first recommendation uses the maturity assessment and *Six Sigma* strategies. The assessment describes nine areas required to achieve SCM maturity. Project efforts focused on the lowest of these areas to establish base procedures. *Six Sigma* implies that customers drive the business. The combination of the COPIS (Customer–Output–Process–Input–Supplier) model and self-assessment indicated that the greatest advance in ITIL SCM maturity can be attained by tracking customer satisfaction, and using that information to proceed with the other levels.

A second recommendation is to implement a standard SLA process. A process map provided a foundation for end-user problem handling. Most end users have little interest in any process not affecting them. Aligning SLAs to the technical handling of problems and solutions via a helpdesk provides another method to identify key maturity objectives. The advantage to taking this step is that it provides a method to compare helpdesk calls with other forms of user feedback.

The final recommendation derives from an overall project goal. Due to SCM immaturity, financial constraints were of little interest in comparison to process development. Alignment with ABC allows resources to be used more effectively. Integration of ABC will evaluate and prioritize activities performed as per the first two recommendations and in turn link SCM with FM. SCM maturity relies on client satisfaction and the number of reasonable approaches to achieving maturity is numerous with project recommendations providing maturity growth potential since they mandate solid customer relationships. Successfully linking SC, SLM, AM, CM, and FM is integral to comprehensive ITIL maturity and reflective of IT excellence.

11.11 Conclusion

Focusing on the customer has a trickle-down effect. To better serve the constituents of DC, OCTO determined that a consolidated IT infrastructure would be more efficient. Converging databases yields less data redundancy and more accurate outputs so that OCTO will converge 95% of all the District's databases into two central databases that communicate through a central WAN. Consolidation and centralization of data communications will place greater emphasis on IT continuity. The entire IT infrastructure must be backed up and recovery efforts initiated almost immediately when failures occur. In addition, IT continuity will ensure contingency plans in case of emergencies. Centralized IT infrastructures such as the one that OCTO will implement depend on IT continuity. The level of efficiency required by agencies must consistently be met regardless of conditions. This commitment can only be achieved by adherence to the most efficient IT and quality practices, in the end transforming Washington, DC into the *City of Access*.

References

Edgeman, R. and Bigio, D. (2004). Six Sigma as Metaphor: Heresy or Holy Writ? *Quality Progress*, 37(1), 25–30.

Edgeman, R., Dahlgaard, S.M.P., Dahlgaard, J. and Scherer, F. (1999). Leadership, Business Excellence models and core value deployment. *Quality Progress*, 32(10), 49–54.

Klefsjö, B., Wiklund, H. and Edgeman, R. (2001). Six Sigma seen as a methodology for Total Quality Management. *Measuring Business Excellence*, 5(2), 31–35.

Part III

Applications of Six Sigma in Transactional Environments

12

Increasing newspaper accuracy using Six Sigma methodology

Ronald D. Snee

12.1 Introduction

Shortly after taking over the leadership of the newsroom of a major newspaper, the new editor initiated an effort to improve the quality of the newspaper. One of the first questions leaders should ask is 'Is this problem important?' The answer here was a clear yes. Nothing is more important to a newspaper than the accuracy of the names, facts, figures, and other information it publishes. If the facts are wrong, a name is misspelled, or a mathematical mistake is made, the newspaper looses the reader's trust and the newspaper's credibility is reduced. For example, this newspaper had reported the promotion of a new CEO in a major US corporation and misspelled the new executive's name. The newspaper received a call from an unhappy reader – not the new CEO, not the company's public relations department, not the CEO's spouse, but the CEO's mother! This points out that the requirement of accurate information can come from many different sources, some of which may not be anticipated.

Some benchmarking was done early in the project to determine what other newspapers had done to eliminate errors. Some newspapers had taken a punitive approach and penalized reporters for mistakes. Others tried giving employees bonuses for keeping errors low versus other departments. In some instances this competition created antagonism between individuals and departments. It was decided to find ways to improve the story-writing process in a way that enabled reporters and others to reduce errors. The following discussion is an elaboration of the 'Newspaper Accuracy' case from Hoerl and Snee (2002). While this case is about reducing errors in newspaper publishing, it is illustrative of reducing errors in processes in general.

The Six Sigma improvement framework and tools were used to guide the work of the improvement team. The Define, Measure, Analyze, Improve, Control (DMAIC) phases are discussed below with emphasis on the tools used in each phase, the purpose of each tool, and what was learned from the application of each tool. A brief description of the most common Six Sigma improvement tools is given in Table 12.1. A depiction of the purpose, deliverables, and key tools of each step in the DMAIC process is shown in Figure 12.1.

A review of the Six Sigma tools in Table 12.1 shows that some, but not all, involve statistical tools; namely the process capability, measurement system

Table 12.1 Key Six Sigma tools

Process map	A schematic of a process showing process inputs, steps and outputs
C&E matrix	A prioritization matrix that enables you to select those process input variables ('X's) that have the greatest effect on the process output variables
Measurement system analysis	Study of the measurement system typically using Gage R&R studies to quantify the measurement repeatability and reproducibility
Capability study	Analysis of process variation versus process specifications to assess the ability of the process to meet the specifications
FMEA	Analytical approach for identifying process problems by prioritizing failure modes, their causes and process improvements
Multi-vari study	A study that samples the process as it operates and by statistical and graphical analysis identifies the important controlled and uncontrolled (noise) variables
DOE	A method of experimentation that identifies, with minimum testing, how key process input variables affect the output of the process
Control plan	A document that summarizes the results of a Six Sigma project and aids the operator in controlling the process

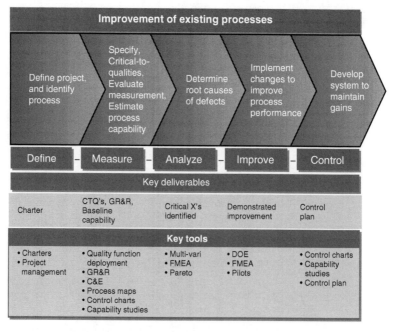

Figure 12.1 The DMAIC methodology and key tools.

analysis, Multi-vari studies, and design of experiments (DOE) tools. The other tools, process mapping, cause and effect (C&E) matrix, failure modes and effects analysis (FMEA), and control plans, are process focused and utilize statistical thinking while not involving statistical tools directly.

12.2 Define phase

In the define phase you select a project to work on, and define the specific problem to be solved and process to be improved. Key process metrics are used to guide project selection and to identify the goals for the project. The resulting project and its objectives are summarized in the project charter. The project charter is a key tool of the project definition phase. A common cause of failure in improvement projects is for different people to have different understandings of what the project will do and accomplish. This can lead to disappointment and finger pointing at the end of the projects. In the newspaper accuracy study the leader (editor) established error reduction as an important issue. In order for a Black Belt (BB) (project leader) to work on this issue we need a project charter that defines the work to be done including the process involved, problem statement, the metrics (process baseline, project goal, process entitlement) that are associated with the problem, the project objective, the financial impact of the project, the team members, and project scope.

Some baseline data showed that while the copy desk could catch and fix as many as 30–40 errors per day, the error rate on a typical day was 20 errors. A goal for this project was set to reduce the errors by 50%; that is, to less than 10 errors per day. The financial impact of an error was established as $62 if caught at the copy desk, $88 if caught at the composing room, $768 if a page had to be redone, and $5000 if the presses had to be stopped and restarted. Of course the cost of an error being published is 'unknown and unknowable.'

There were a number of secondary objectives for the project including:

- Elimination of rework and duplicative work throughout the processes used to create stories, graphics, and photos.
- Free the copy desk editors to do value-added work rather than unrewarding spell checking, cross checking, and fact checking.
- Breakdown barriers between departments in the newspaper.
- Reduce costs by reducing the number of killed pages and late night fixes.
- Reduce costs by finding alternated sources of financial information.
- Use information technology to automate chart creation.
- Create a culture that emphasizes 'doing it right the first time' and taking personal responsibility for increasing accuracy.

This seems like a long list of objectives. But you will see that fortunately the primary objective of reducing errors by at least 50% and almost all of the secondary objectives were met by this project.

An operational definition for 'an error' was created before any data were collected, so that the data would be accurate, and everyone would be talking about the same thing when errors were discussed. An error was defined as (1) any deviation from truth, accuracy, or widely accepted standards of English usage, or (2) a departure from accepted procedures that causes delay or requires reworking a story or a graphic. It was also decided to divide the errors into nine categories: misspelled words, wrong number, wrong name, bad grammar, libel, word missing, duplicated word, wrong fact, and other.

The 11-person team consisted of the BB, the editor, two copy editors, two graphics editors, one reporter, and four supervisors. This team was large, but was effective. Generally, teams should consist of not more than 4–6 people. Larger teams have difficulty finding mutually agreeable meeting times, and have problems reaching consensus and making decisions. Fortunately, team size was not a problem in this case.

12.3 Measure phase

The measure phase is intended to insure that you are working on improving the right metric; one that is truly in need of improvement and which you can measure well. In the measure phase we select the appropriate process outputs to be improved, based on the objectives of the project and the customer needs. Acceptable performance is determined and you gather baseline data to evaluate current performance. This work includes the evaluation of the performance of the measurement system, as well as the performance of the process being studied. The tools used during the measure phase include the process map, C&E (fishbone) diagram or C&E matrix, measurement system analysis, capability analysis, and a control chart analysis of the baseline data on the process output. Process mapping, C&E diagrams, capability studies, and control charts are popular improvement tools discussed by Hoerl and Snee (2002) and other authors. More detailed discussions of the analysis of measurement systems are found in AIAG (1990). Breyfogle (1999) discusses the use of the C&E matrix.

The process map is prepared by the team, not the BB alone, and provides a picture of the process, as well as enabling the identification of non-value-added work and the 'hidden factory' where the reworking is done. Rework refers to redoing sub-standard work done previously, such as finding and correcting errors in financial reports. Non-value-added work refers to work that adds no value to the product or service, but is currently required due to inefficiencies in the process. For example, warehousing finished goods adds no value from a customer point of view, but some level of warehousing is typically needed to effectively supply customers.

The first process map is prepared at a high level, usually consisting of 5–10 steps. If further detail is needed for a few key steps, these steps are mapped further, creating 5–10 sub-steps. Important process input and output variables are usually identified during the process mapping work. The newspaper writing and editing five-step process map is shown in Figure 12.2. Note the 'revision cycle' which may be a source of both non-value-added work and rework. The size of the paper (number of pages), number of employees absent each day, and a major change in front-page (cover) story (yes/no) were identified by the team as variables that could have an effect on errors.

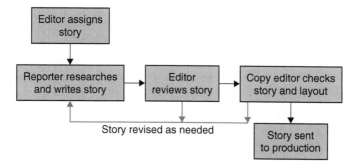

Figure 12.2 Newspaper writing and editing process.

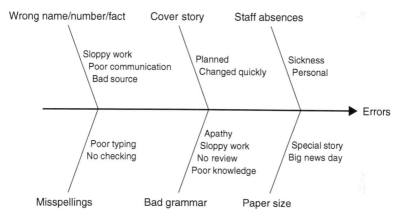

Figure 12.3 Causes of newspaper errors (C&E diagram).

The purpose of both the C&E diagram and C&E matrix is to enable the BB and the team to study the relationships between the process input variables and the process output variables. In the case of the newspaper accuracy study, the key output variable is errors. The C&E diagram serves as a visual display of the 'effect,' or output variable, and its important 'causes' or input/process variables. The C&E diagram for errors is shown in Figure 12.3. The C&E matrix rates the process input variables in terms of their relative impact on the process output variables.

In this project it was decided that the C&E diagram was adequate and that the C&E matrix was not needed. As a general rule, one does not have to use every tool on every project; rather we use whatever tools are needed to successfully complete each phase of the DMAIC methodology. Note also that the process map, C&E diagram, and C&E matrix are examples of 'knowledge-based' tools, that is, they are developed based on our existing knowledge of the process, rather than objective data. Eventually, we need objective data to ensure our current understanding is correct, and to enhance this understanding.

In this case the measurement system analysis consisted of developing the measurement system and errors collection scheme, and validating it. In other instances, particularly in manufacturing, Gage repeatability and reproducibility (R&R) studies (AIAG, 1990) are used to evaluate the adequacy of the measurement system. Gage R&R evaluate our ability to replicate results when we take multiple measurements (repeatability), and the ability of several people or pieces of measurement equipment to obtain similar measurements (reproducibility). Analysis of variance techniques are used to quantify the amount of variation due to repeatability and reproducibility.

Measurement system analysis is a particularly important step because, often as much as 50% of the measurement systems in use are in significant need of improvement. Of course, there can be other measurement issues besides repeatability and reproducibility, such as accuracy (ability to achieve the correct average measurement). In Six Sigma projects outside of manufacturing, such as the newspaper study or in finance, most of the measurement system work focuses on the creation of the measurement system and the construction of the data collection process.

A process capability study is conducted to measure how well the process is capable of meeting the customer specifications. Typical outputs of these studies are short-term capability indices (short-term capability level, Cp, Cpk) in the measure phase and long-term performance indices (long-term capability level, Pp, Ppk) in the control phase. Such studies are often done using control charts (Montgomery, 2001). A control chart analysis of 44 days of baseline data showed that the errors were being produced by a stable process with an average value of approximately 20 errors per day, with daily variations from just below 10 to just below 40. A control chart is a plot of data over time with statistically determined limits of 'normal' variation. A run chart is analogous to a control chart, but does not have the statistically determined limits. Figure 12.4 shows a run chart of this data illustrating the degree of stability, average level, and variation of the process.

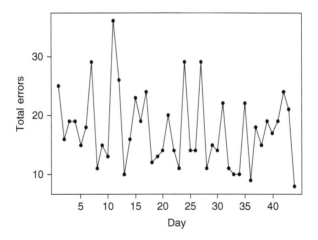

Figure 12.4 Newspaper errors, March–April.

12.4 Analyze phase

The analyze phase helps us avoid the 'ready, fire, aim' approach by accurately diagnosing the root causes of problems. In the analyze phase we evaluate the baseline data to further document current performance and to identify root causes of the observed variation and defects. Additional data are collected as needed. Two of the most commonly used improvement tools in the analyze phase are Multi-vari studies and FMEA. Multi-vari studies are process studies in which we collect data on the key process and input variables as well as the key outputs. The data are then analyzed using graphical and statistical tools (e.g., regression analysis, hypothesis testing, etc.) to identify the variables having the most significant impact on the output variables. FMEA is a disciplined methodology for identifying potential process defects, and taking proactive steps to avoid them (AIAG, 1995).

When data on errors are evaluated a Pareto analysis is often used to determine which categories of errors are causing the major errors. The Pareto chart (see Figure 12.5) is basically a bar graph where the bars are ordered by number or magnitude of occurrence. Knowledge of the important error categories will suggest root causes of errors. The theory is that a few categories will account for the majority of the errors. This theory holds in the newspaper errors study. In Figure 12.5 we see that the majority of the errors during the March–April time period are due to misspelling, wrong names, numbers and facts, and poor grammar.

Attention was first focused on the categories producing the largest number of errors: spelling, wrong names, facts and numbers, and bad grammar. One root cause identified was that the reporters were not using the spell checker. The attitude was 'I don't have time to spell check. Besides, the copy editors will catch the errors any way.' The reporters were also not routinely checking their facts and their sources, which was a job requirement. We address how to deal with these root causes in the improve phase.

A Multi-vari study is conducted to identify variables that may be producing the errors. The variables studied were those identified in the measure phase;

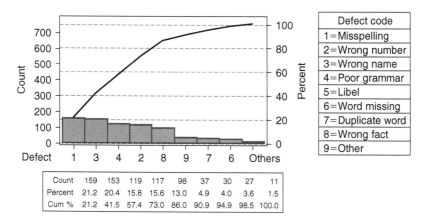

Figure 12.5 Pareto chart for errors.

namely, size of paper, number of employees absent, and major changes in the front-page story. While the size of the paper is controllable, the other two variables are not controllable. Day of the week (Monday, Tuesday, Wednesday, Thursday, and Friday) and month of the year (March–December) differences were also studied. Day-of-week and month-to-month differences, if present, provide clues to other sources of errors. Work teams often perform differently on Mondays and Fridays than on the other days of the weeks. Analysis of the data indicated that the size of the paper (more pages leads to more errors) and changing the front cover story (new stories had to be created under very tight schedules increasing the error rate) had an effect on errors.

The Multi-vari study also detected a drop in errors in the early part of the study even before any improvements had been made. This change was attributed to the 'Hawthorne effect' (named after the Hawthorne plant in which the phenomenon was discovered) in which improved performance results from management attention rather than any real change in the process. Unfortunately we don't know what caused the improved performance or how to sustain it. Worse yet, the improvement will go away when management attention is reduced. Juran and Gryna (1988, pp. 10.12–10.13) provide more discussion of the Hawthorne effect.

12.5 Improve phase

In the improve phase we figure out how to change the process to address the root causes identified in the analyze phase, and thereby improve the performance of the process. Each process change is tested with a confirmatory study to verify that the predicted improvements in fact happen. Several rounds of improvements may be required to reach the desired level of performance. Note that this is the only phase in the DMAIC process that actually makes improvement. The other phases are intended to properly set up (DMA) and maintain (C) the improvements from this phase.

In the newspaper case, it was reaffirmed that reporters had the responsibility to check the accuracy of their articles. Reporters are required to verify the correctness of every important element of the story or graphic. This was not being routinely done previously. After this requirement was put in place more than 90% of the facts are verified before they reach the copy desk. This means that copy editors can do meaningful work not just check facts. One copy editor used to call the library 25–30 times a day to check facts. That dropped to less than five calls per day.

Three job aids were also created: a 'Spell Check How-To,' a list of 'Ten Rules of Grammar,' and the 'Pyramid of Trust' which detailed the sources that could be trusted to produce accurate names, facts, and numbers. Reporters were also taught how to avoid common mathematical mistakes. These new working methods were communicated in an 'All-Hands' meeting in July. The importance of being careful when the front cover story changed with little notice and publication of large size editions of the newspaper were also discussed. The interim goal of producing less than 10 errors per day reaching the copy desk was also reviewed and reaffirmed.

It was now time to test whether the changes were in fact having an effect. One month went by and the data for the month of August were analyzed. The result was that total errors had not changed! The management team was assembled and the situation reviewed. Why were the errors still high? It was learned that the new procedures were not being used. Many employees did not feel that management was serious about the changes, and therefore did not take the changes seriously. This emphasizes the point that deciding on improvements, and actually implementing them effectively, is two different things! The Editor reiterated that the new procedures were to be used and that the management team was expected to lead this new way of working. Another 'All-Hands' meeting was held to address the issue.

One month later when the latest data were analyzed the total errors had dropped significantly (Figure 12.6). In another month the total errors had dropped by approximately 65% as compared to the goal of 50% reduction! The new procedures were clearly working. Finding that new procedures are not being used is not an uncommon occurrence. It is leadership's responsibility to ensure that the new way of working is used. Otherwise, the benefits of the project will not be realized. Project reviews, confirmatory studies, and process audits are effective ways to identify whether the process changes are being used and are effective.

In some instances, particularly in manufacturing processes, additional work needs to be done in the improve phase to get the desired process performance. Typical studies include verifying C&E relationships identified in the measure and analyze phases, identifying optimum operating conditions, and defining process specification limits. The tools of statistical DOE and response surface methodology can be very helpful in these instances (Box *et al.*, 1978). Designed experiments have been successfully used in non-manufacturing studies and their use continues to grow in this important area of improvement (Koselka, 1996).

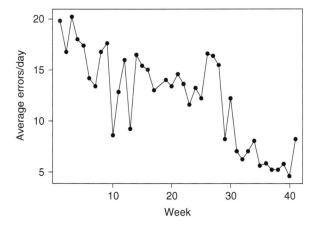

Figure 12.6 Weekly average newspaper errors March–December.

12.6 Control phase

In the control phase we install a system that will insure that the improved performance of the process is sustained after the Six Sigma team has completed its work and moved onto another project. The key tools of this phase are the control plan (AIAG, 1994) including control charts and long-term capability studies.

Returning to our newspaper case, we see that while the errors were significantly reduced, errors were not at zero yet, and more work was needed to achieve further improvement. In the meantime, a control plan was put in place to hold the gains of the work done to date, and to keep the errors at the level obtained. This is the purpose of the 'control' phase in Six Sigma, to hold the gains. The control plan specified that the following measures would be monitored using control charts:

- total errors,
- errors by category,
- percentage of articles checked by the author,
- percentage of articles spell checked.

The latter two measurements were found to be particularly useful in detecting when the reporters were not following procedures and thus errors would be a problem.

Checklists were also created and roles and responsibilities, including backups, were defined to reduce handoff problems between departments. This enabled people to view their work processes as part of an overall system. Work was also initiated in finding the sources of errors in the graphics.

Future projects are typically identified at the conclusion of the control phase. In this case it was determined that errors could be further reduced by improving the process by which the graphics in the newspaper were produced. A new team was created to work on this opportunity.

12.7 Results

This work resulted in the errors being reduced 65% producing a savings in time at the copy desk of more than $226,000 per year. This was a cost avoidance rather than a hard dollar savings as the persons involved were freed up for other more value-added work and did not leave the employment rolls. A modest hard dollar savings of approximately $30,000 per year did result due to the cancellation of a contract resulting from the findings of this project. The actual savings were even more because of the effects of fewer errors in the composing room, fewer pages needing to be redone and fewer press stops. These additional benefits were not quantified.

Other benefits resulted from the new way of operating, including:

- Fewer missed deadlines, including the ability to deal effectively with extremely tight deadlines. In the case of one major news story, the newspaper was closed 26 min early which the copy desk says would have been impossible without using the new methods.

- Improved morale at the copy desk. Copy editors were freed up to make better use of their talents and training. One copy editor commented 'I feel like I'm using my 10 years of journalism experience, rather than spending my time typing messages to the library, which a high school student can do. Now copy editors are freed to use much more of their talent to add value to the newspaper.'
- Re-keying of names (rework) reduced. In some instances a name was re-keyed six times before it appeared in the newspaper. The team found a way to reduce it once thereby eliminating five opportunities for error.
- More efficient and less costly sources of information were found, resulting in reduced errors and less number input time (less keying of data). News assistants were freed up to do more valuable work.

Fewer errors resulted in less non-value-added work and a more streamlined and effective process. Such effects are characteristic of what happens when the errors and defect levels of business processes are reduced. The processes work more effectively and efficiently, costs reduce, employee morale improves, and customer satisfaction increases.

It is also noted that two commonly applied improvement tools, FMEA and DOE, were not used in this newspaper accuracy project. This is not unusual. In any given project, some tools are not needed, as was the case in the newspaper accuracy project. In other cases, the work may have already been done in a previous project, such as the process map and measurement system analysis. Sometimes the unused tools will be needed in subsequent studies. In the newspaper accuracy case the error rate was reduced to less than 10 per day, which is not zero, the desired state. FMEA and DOE may be useful in subsequent projects intended to make further improvements.

12.8 Success factors and lessons learned

A formula for success of any project is support and involvement of leadership as well as working on the 'right project,' involving the 'right people' and using the 'right methods and tools.' All of these success factors were present in this project. The support and involvement of senior management was present because the problem was important to the success of the newspaper. Accuracy is a core value of all newspapers. The team consisted of the right people, those who were involved in the story writing process and had ideas on how to improve it as well as the dedication and will to pursue the improvements. In addition to the proven DMAIC improvement framework the team created an operational definition of what an error was and then instituted a data collection system to monitor errors and find ways to reduce errors. Collection and analysis of data was new to this organization. They also displayed in the newsroom graphs and charts of their work communicating to the organization the purpose and focus of their work and the progress that was being made.

The team had several key learnings from this work. First they learned that the DMAIC improvement framework works if you follow the process. If you don't follow the process (e.g., take short cuts, skip steps, etc.) the success rate drops off quickly. They also learned that when management provides leadership; identifying important problems and needed improvements based on data, the organization will adopt the improvement with improved performance resulting. People need to know that leaders know where they are headed and are serious about the stated direction. The team also learned the power of communication with the organization using charts and graphs.

12.9 Deploying Six Sigma company-wide

The newspaper errors project illustrates what goes on in a single Six Sigma project. In an organization that is utilizing Six Sigma many projects are undertaken and completed with full-time BBs completing 3–5 projects each year and Green Belts (GB) who work part time completing 1–2 projects each year. As soon as a project is completed the BB or GB selects another project from the 'project hopper' – the list of projects selected and prioritized and ready to go – and begins the improvement work. Six Sigma projects are conducted in all parts of the organization. To get a sense of the size of such an activity a rule-of-thumb is an organization should have 1 BB per every 50 professional employees and 10 GB per every BB. These ratios of course vary from organization to organization. The point is that the number of BBs, GBs, and Six Sigma projects can be quite large in large organizations. For example Bank of America recently reported that in a little over 2 years they have trained more that 10,000 Champions, BBs, GBs, and Master Black Belts (MBBs) which have produced more than $2 billion in savings (Jones, 2004). The Six Sigma approach provides the strategy, methodology, and tools for leading and managing such a large improvement effort. Additional discussion of deployment of Six Sigma in an organization including success factors, needed infrastructure, roles, tools, and deployment roadmaps are contained in Snee and Hoerl (2003; 2005).

References

Automotive Industry Action Group (1990). *Measurement Systems Analysis Reference Manual.* Southfield: AIAG.

Automotive Industry Action Group (1994). *Advanced Quality Planning and Control Plans.* Southfield: AIAG.

Automotive Industry Action Group (1995). *Potential Failure Mode and Effects Analysis Reference Manual* (2nd Edition). Southfield: AIAG.

Box, G.E.P., Hunter, W.G. and Hunter, J.S. (1978). *Statistics for Experimenters.* New York, NY: John Wiley and Sons.

Breyfogle, F.W. (1999). *Implementing Six Sigma – Smarter Solutions Using Statistical Methods.* New York, NY: John Wiley and Sons, pp. 277–280.

Hoerl, R.W and Snee, R.D. (2002). *Statistical Thinking: Improving Business Performance.* Pacific Grove, CA: Duxbury Press.

Jones Jr., M.H. (2004). 'Six Sigma … at a bank', *Six Sigma Forum Magazine*, February 2004, 13–17.

Juran, J.M. and Gryna, F.M. (1988). *Juran's Quality Handbook* (4th Edition). New York, NY: McGraw-Hill.

Koselka, R. (1996). The new mantra: MVT (multivariable testing). New York, NY: *Forbes*, March 11, 1996, 114–118.

Montgomery, D.C. (2001). *Statistical Quality Control* (3rd Edition). New York, NY: John Wiley and Sons.

Snee, R.D. and Hoerl, R.W. (2003). *Leading Six Sigma – A Step-by-Step Guide Based on Experience at GE and Other Six Sigma Companies*, Upper Saddle River, NJ: Financial Times Prentice Hall.

Snee, R.D. and Hoerl, R.W. (2005). *Six Sigma Beyond the Factory Floor – Deployment Strategies for Financial Services, Health Care, and the Rest of the Real Economy*, Upper Saddle River, NJ: Financial Times Prentice Hall.

13

An application of Six Sigma in human resources

Alan Harrison

13.1 Introduction

This case study describes the application of Six Sigma, DMAIC (Define–Measure–Analyze–Improve–Control) approach in human resources (HR) function and processes within an engineering organization that serves over 1000 employees. It is a traditional, 'brown site,' engineering company, organized in business divisions and central support functions, part of a global group with over 20 different worldwide locations.

The two main drivers for adoption of Six Sigma in HR were at organizational and HR functional level.

At organizational level – the company ownership had just been changed and corporate worldwide integration process was in its early stage. The board defined and communicated the new vision of integrated global company, strategic objectives and emphasized the need to bring together different business cultures, procedures and effectively recognize and share best practices across the global organization. Six Sigma was selected as the corporate business improvement strategy and approach, encompassing all employees in all business processes.

At a functional level – the newly appointed HR director defined and communicated his vision and core values to the HR team, and that Six Sigma was to become a way to make shared vision become reality. The author of this case study facilitated HR team in 'from vision to reality' journey.

The project team was formed that consisted of the HR team, Six Sigma Black Belt – facilitator and HR director – the project mentor. Internal customers were involved in all the project steps.

Key phases of the journey were definition of the ultimate goal, that is a vision of the 'new world,' selection, motivation and training of the HR project team and deployment of Six Sigma DMAIC model to define and drive achievement of improvements and to sustain continuous improvement.

HR Vision was defined and shared in a kick-off workshop as:

Right People in the Right Place at the Right Time at the Right Cost

This means sourcing, selecting and motivating people that meet requirements of a particular job function or tasks, just-in-time when required and at or better

	Traditional	Six Sigma
Manufacturability	Trial & error	Robust design
Analysis	Experience	Data
Focus	Product	Process
Time	Reactive	Proactive
Planning	Short term	Long term
Control	External	Self-control
Psychological contract	Compliance	Commitment
Structures & systems	Bureaucratic	Organic
Employee relations	Low trust	High trust
	Personnel	HRM

Figure 13.1 Core values comparison.

than budgeted cost of employment to maintain overall business competitiveness and structure. Top-level objectives were agreed: to enjoy improvements, to continuously communicate and to recognize and satisfy internal customers and their real needs.

A comparison between Six Sigma and HR Management (HRM) core values was discussed, which revealed that they both share the same socio-technical values, as presented in Figure 13.1.

For example, HRM's 'high trust' is also a vital element of every Six Sigma DMAIC step:

- High trust underpins teamwork and prioritization and sharing of problem definition.
- High trust is required for collection of complete, honest, accurate, relevant and timely (CHART) measurement data.
- High trust is required to get all team members to accept and believe analysis results.
- High trust stimulates right mind-set to generate new, 'out-of-box' ideas for improvement, with no fear at individual level.
- High trust is essential to empower process participants to control and sustain their part of an improved process.

This analysis and use of relevant HR terminology helped HR team to start understanding and accepting Six Sigma as 'much closer to home than initially thought,' not just 'some statistics used in manufacturing.' The end result of the HR team journey is better, faster and more cost-effective HR service to the organization, improved internal-customers' satisfaction, improved motivation and job satisfaction of HR team, all contributing toward better business performance.

13.2 Before improvements

Before improvements, the HR function was seen as reactive, over-manned, unprofessional, delivering low quality, slow and cost non-effective service. HR employees were working hard and suffered day-to-day frustrations from running broken and non-effective processes that had no clear outputs. Performance monitoring was ineffective and inefficient, for example a lot of effort was put into producing half-an-inch thick monthly HR report that would usually get dumped unread into a bin by business division's heads.

The high level of HR team effort was achieving a fraction of required results, which had a demoralizing effect on the team whose contribution was not appreciated in the business. Negative feedback was acting as a negative control loop, which also had a negative impact on almost all employees, as HR was not always there to help employees to resolve their personal or business related problems.

Many research studies have indicated that HRM processes have high impact on business performance. For example, in their 10-year study of hundred UK small and medium size businesses, Patterson *et al.* concluded that:

> HRM practices…are the most powerful predictors of change in company performance

Fixing HR practices was recognized by the board as a vital element in delivering global integration and business objectives.

A surgical cut was required to break this negative 'catch 22' loop. The new HR director and Six Sigma facilitators prepared a 'kick-off' meeting. The vision was developed and shared and the team was pulled into the new direction through immediate but focused improvement actions. Six Sigma facilitator moved into the HR department, and stayed there for a number of months, whilst still running other improvement projects in different area of business. Co-location enabled continuous communication, just-in-time training and coaching and immediate actions on observed behavior.

13.3 Define phase

13.3.1 People

Improvement team consisted of processes participants, facilitator (Six Sigma Black Belt) and mentor (HR director). Internal customers were identified as critical stakeholders and they participated through direct representation or two-way communication. Table 13.1 presents the roles and responsibilities of team members.

Project facilitator and mentor repeatedly emphasized and insisted that every participant has to be upfront clear on their and other team members' objectives, tasks, roles and responsibilities and set target dates.

13.3.2 Processes

The first challenge for the HR Team was to recognize their internal customers and understand the basic of Six Sigma – customer needs – suppliers action relationship. It was essential to get this mind-set as the right 'soil' for other Six Sigma 'seeds' that were to follow, for example a process way of thinking, DMAIC,

Table 13.1 Main goals, roles and responsibilities of the team members

Stakeholder	Goal	Roles and responsibilities
Internal customers	Alignment with the business objectives and strategy	• To recognize internal suppliers and their processes • Participate in definition of business needs, criteria for improvements' definition and prioritization and criteria for selection of KPIs • Complete, timely and positive feedback and contribution
HR process owners and participants	To deliver satisfaction of all stakeholders	• To recognize internal customers and their needs • To completely own their processes • To help internal customers in their roles and responsibilities, as process experts • To learn and use right Six Sigma improvement tools • To develop, implement and sustain in continuous improvement of their processes
Six Sigma Black Belt	To achieve deployment of right improvement tools that deliver required improvements in a most effective and efficient way	• To build a team • To lead/coach clarification of objectives, timescale and resources • To identify and provide Six Sigma training, as required by delivery of specific improvement tasks (just-in-time Six Sigma training) • To facilitate and coach team in making Six Sigma become a 'way-of-life'
Mentor	To create environment that supports and continually drives required improvements	• To define and share vision of the 'new world' and team core values, based on integrity, honesty, openness and mutual respect to each other • To gain commitment and trust, select, motivate and lead the team • To help effective and efficient decision-making • To eliminate or resolve barriers to progress

supplier–input–process–output–customer (SIPOC), process mapping, facts-based decision-making…

The second radical view was to re-define HR as six basic processes with a purpose to provide outputs that satisfy internal-customers' needs:

1. Organization development.
2. Employee development.
3. Resourcing.
4. Reward.
5. Communication.
6. Organization improvement.

A model was developed to help the team visualize the relationship among those processes, and develop process way of thinking, as presented in Figure 13.2. Organization development and organization improvement encompass the key four processes. All processes are owned by suppliers and customers.

Figure 13.2 HR model.

The first breakthrough came when team started to comprehend process thinking in a service industry. A number of 'ordinary life' examples (e.g., buying house insurance, getting a hair cut...) were discussed and mapped prior to transferring the same approach to professional services, including specific HR processes.

The facilitator continuously promoted a process way of thinking by repeatedly asking the same questions, for example 'Who is the customer? What do they really need? How do you know that? Where does this go to and come from?' even when preparing a coffee or tea during the break. This did cause some laugh and jokes but made customer's needs and process way of thinking become every day's terminology.

13.3.3 Definition of customers' needs and objectives

SIPOC was used to identify internal customers and capture their needs against all HR processes. Information was collected through a series of structured interviews and brainstorming.

Simplified quality function deployment (QFD) approach was used to analyze captured internal-customers' needs – 'Whats,' 'Hows' were used to identify and prioritize value-added HR processes or sub-processes. This visual, matrix presentation made invaluable contribution in aligning discussions within the team and with internal customers, achieving sharing and prioritization of goals and objectives against identified HR processes.

An illustrative example for the recruitment process matrix is presented in Figure 13.3. This approach helps requirement management and trade-offs, for example 'advertising' step may have a considerable impact on attracting right people and can discourage selection of cheapest job advertising method.

Team agreed on the following top-level objectives:

- Improve and implement HR processes and key performance indicators (KPIs).
- Develop and embed continuous improvement.
- Increase job security and survival of HR central function.

What	Process requisition	Advertise	Select	Obtain acceptance	Inform unsuccessful
Position filled					
Right people		▲	●	▲	
Right time	●	●	▲		○
Right place		▲	▲		
Right cost	●	▲		●	
Targets (weeks)	1	2	3	4	5

● Strong
▲ Medium
○ Weak

Figure 13.3 Simplified QFD – recruitment process.

- Promote Six Sigma in non-manufacturing areas.
- Promote empowered, high performance teams.

13.4 Measure phase

The role of measurement is to trigger new questions, as same questions most likely produce same results. As we want to change results we need to ask different questions that will open new doors and point into the new directions of breakthrough improvements.

13.4.1 Key performance indicators

One of the initial barriers was a presence of a mind-set within the team that 'HR cannot be measured, it is not a manufacturing.' This issue was tackled step-by-step, not 'head-on,' where facilitator led and helped the team to realize that, actually, HR can be measured as well. The first step was to get team's understanding of internal customer–supplier link. Follow on was to identify customer, their needs and ask the question 'What makes a logical sense to measure and does link to a particular customer need?'

After this, a KPI would become understood by the team, for example how a 'return rate of feedback' is relevant metric to communication process, or 'throughput time' is one of resourcing process metrics.

A generic criteria for the selection of KPIs was agreed by the team, with input from internal customers. The focus was on balancing quality–cost–delivery–behavior, as follows:

- Alignment between HR and business strategic objectives.
- Focus on value-added to organization or an individual.

- Critical-to-quality measures (CTQ).
- Critical-to-cost measures (CTC).
- Critical-to-throughput time measures (CTT).
- Positive impact on behavior, where applicable.
- Easy to understand and remember.

For example, in Resourcing process, recruitment and transfer time (CTT) were balanced against budget (CTC) and turnover (CTQ). Communication, resourcing, reward and development, the core processes in developed HR model (presented in Figure 13.2), were selected as first processes to be measured and analyzed.

Process mapping was used extensively and proved to be one of most effective analysis and improvement tools. 'As-is' process maps were created and analyzed the same way as any manufacturing process. A process would have been walked and a standard set of questions was developed and used to interview internal customers and other participants in order to trigger and capture their comments related to 'as-is' and ideas for 'to-be' process.

The key benefit of process mapping was an effective way that revealed shortcomings with obvious and immediate remedies, for example:

- Identification of 'hanging-in-the-air' steps with broken flow that had to be either eliminated or connected.
- 'Looped' steps that looked like HR – business units were playing ping-pong, for example in Recruitment process, a requisition form would have bounced four to five times before proceeding to the next value-added step. The obvious solution was to simplify and re-design requisition form with clear agreement and indication who needs to complete what fields.

Typical cycle time and overall throughput time were estimated or measured. A number of concerns were captured during the 'process-mapping walk,' those were analyzed and improvement actions were agreed and monitored. Too many tasks had ambiguous ownership and RACI (responsible–accountable–consulted–informed) was used to capture and clearly agree the roles and responsibilities against every activity on a process map.

13.4.2 Cost of poor quality

Cost of poor quality against each process was defined and estimated. Some costs are tangible and measurable, for example costs of involvement in employment tribunals, which is a possible outcome of poor quality of redundancy or resourcing processes. Most of HR processes have less tangible, but potentially higher cost of poor quality, for example poor quality of communication process could cause various and very visible problems. The team realized that there was a lot of perception involved and that it usually takes more effort and time to change already created perception than to create a new one.

A priority was given to simplification and clarification of processes, in particular information flow, scope and KPIs that were within the control of the HR team.

13.5 Analyze phase

Prior to commencing any sort of analysis, even discussion of findings, all data were tested against the CHART data test. Six Sigma is a data driven methodology, which utilizes a number of analysis tools. In order to get the best out of those tools and prevent waste of time and effort or even risk of losing credibility of the approach, an analyst has to make sure that the input data is of satisfactory quality.

The primary goal of analysis phase was to identify vital few parameters that control process performance and results. The secondary goal was to test and prove Six Sigma approach and tools in a non-manufacturing process. Process owners performed analysis. Black Belt facilitated analysis and implemented just-in-time training, that is prior to application of a particular improvement tool Black Belt would present a theory of that tool and a couple of examples of application and results achieved in his previous projects. Only then would he ask a team member(s) to perform agreed analysis and think about improvement actions.

A number of simple but effective Six Sigma tools were deployed to present and analyze KPIs and performance of HR processes. Whenever possible, appropriate graphical analysis tools had to be used as the first analysis step. The team found enlightening and satisfying experience of producing and presenting the charts. For example, run-charts were used to present and understand what had been happening, possible trends, outliers and possible correlation, repetitive or special causes trends (at a run-chart analysis most of those are hypothesis, not necessarily conclusions).

Histograms were used to understand performance against targets and tolerances, if there were any. If not targets and tolerances would be defined based on customer requirements and sometimes based on current process performance. Histograms were in particular useful to visualize and support discussions related to a range and a shape of spread, communicating the message that any process has to be not only on target but has to have a consistent performance as well.

On a few occasions, histograms generated new questions that did lead to the right answers and improvement actions. Statistical control charts were used to visually present and analyze process capability vs. customer requirements, identify, understand and eliminate outliers and to set improvement targets and tolerances.

An example of a control chart, used to monitor communication process feedback, is presented in Figure 13.4.

Although the average result of 3.54 is above the set target (3.1), the lower control limit (LCL = 2.8) is below the target (3.1). In statistical terms, we had a likelihood of not meeting the target.

Histogram and run-chart were used to analyze throughput time of the recruitment process. A few data were identified as outliers, which was confirmed by adding upper control limit (UCL) on a run chart. That chart and a histogram helped recruitment process owners to understand performance of the process they have been living in for a long time.

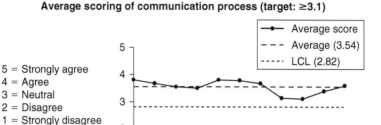

Figure 13.4 Communication process feedback.

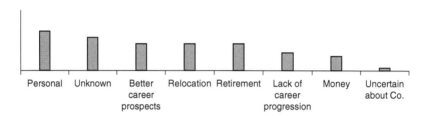

Figure 13.5 Pareto analysis – reasons for leaving the company.

Pareto analysis was used to analyze reasons for leaving the company, as presented in Figure 13.5.

13.6 Improve phase

After understanding the required outputs and targets, vital few inputs and causes of out-of-spec performance, the team agreed specific improvement actions against each HR process. For example, the outliers were analyzed and causes of so unusually long recruitment time in the Recruitment process were identified using cause-and-effect, brainstorming team expert knowledge of specific cases. Key patterns and parameters that have an impact on recruitment time were identified, for example position type, geographical proximity of recruits, type and timing of advertising, competitive position of job package. The team discussed correlation results and agreed balance scorecard in order to prevent sub-optimization of the recruitment process against just one measure.

This resulted in modification of the recruitment methods to prevent future delays. The process performance continued to be monitored using run-charts and Sigma balance scorecard (an example is presented in control section).

Another example is use of cause-and-effect diagram and focused interviews to identify vital few parameters that control feedback return rate. Possible causes of 'communication feedback rate' effect were brainstormed, grouped and prioritized. Communication process was modified in a controlled manner

Table 13.2 Example of improvement and control actions summary

Process	Improvement/control action	Target date	Resp.	Status
Resourcing	Implement Sickness Absence Policy			New
	Amend process			Ongoing
	Analyze exit interview reports to establish common causes			Done
	Introduce a time limit for a decision on internal recruitment			
	Quantify redeployment that has reduced redundancies			
	<other HR processes, actions, responsibilities, dates...>			

Summary of actions to date: Outstanding: # New: ## Completed: ## Total: ##

in order to verify causes of unsatisfactory performance and to get the process under control. A simple example is a briefing communication to communicators, supervisors who cascade team briefings, pointing out when, where and how to communicate and manage feedback communication.

Initial team's concern (mind-set) of whether Six Sigma is applicable to HR was discussed, for example by comparing produced HR control charts and control charts that were produced and used in previous manufacturing improvement projects, for example Communication feedback chart was compared to copper wire thickness chart. This comparison helped the team to understand that manufacturing and non-manufacturing processes or any other process in general share very similar generic pattern, hence Six Sigma can be applied to improve any process.

13.7 Control phase

The key objective of control step is to sustain implemented improvements and embed continuous improvement in the process. Regular team improvement meetings were established. Progress of improvement actions was reviewed in weekly team meetings using a simple improvement and control action summary sheet, extract is presented in Table 13.2.

A database was developed by the facilitator and used by the team to enter and monitor agreed KPIs. Analysis and standard reports were agreed with the team and internal customers and were automatically produced. Any non-standard questions were answered using that database and facilitator was involved in analysis and interpretation of non-standard reports (e.g., correlation analysis).

A quarterly HR review report was developed; containing all HR processes performance results and targets (run charts, control charts, histograms, tables and comments), conclusions, observations, status of improvement actions and sigma benchmark (Z_b) scorecard. Review initially consisted of three A4 pages, later reduced to two pages, compared to the old half an inch thick HR report.

First few reports were taken into business units operations meetings and all details were read out, discussed and explained by dedicated HR team member.

Table 13.3 Summary of Zb scorecard

Line	Process/ sub-process	D	U	OP	TOP	DPU	DPO	DPMO	Shift	Zb
1	Left within 1 year (target = 0)	10	1	54	54	10	0.185	185,185	1.50	2.4
2	Recr. time (target = 5 weeks)	12	1	35	35	12	0.343	342,857	1.50	1.9
3	Transfer time (target = 4 weeks)	3	1	15	15	3	0.200	200,000	1.50	2.3
4	Rejected offers (target = 0)	4	1	70	70	4	0.057	57,143	1.50	3.1
5	Within budget (target variance = 0)	1	1	50	50	1	0.020	20,000	1.50	3.6
	Resourcing	30			224		0.134	133,929	1.50	**2.6**
1	PDPs completion (target = 100%)	30	1	100	100	30	0.300	300,000	1.50	2.0
	Development	30			100		0.300	300,000	1.50	**2.0**
1	Qual. score (target ≥ 3.1)	1.00	1	200	200	1	0.005	5000	1.50	4.1
	Communication	1.00			200		0.005	5000	1.50	**4.1**

Feedback was collected, and further improvements to a report content and format were completed. Very soon operations managers were able to themselves read and interpret results, charts and tables presented in the quarterly review.

Six Sigma Zb balance scorecard was used to summarize reports against each of six HR, example presented in Table 13.3. The advantage of Zb scorecard is its possibility to mix up continuous data (e.g., 'recruitment time' with discrete data; e.g., 'left within year').

Terminology used in Table 13.3 is: defects, units, opportunities for defect, total number of opportunities for defect, defects per unit, defects per opportunity, defects per million opportunities, standard sigma benchmark shift (1.50) and Zb is sigma benchmark.

13.8 Achieved benefits

The number of defects, level of re-work and throughput time have improved in all processes. This was verified by charting selected KPIs and Zb. An example is presented in recruitment process control chart in Figure 13.6.

Tangible cost of HR function per employee has been reduced by 34% over a period of 18 months, with the same or better service provided, through staff re-deployment or nutrition. Businesses have achieved budgeted turnover with 15% fewer people, and, for the first time, annual employee bonus.

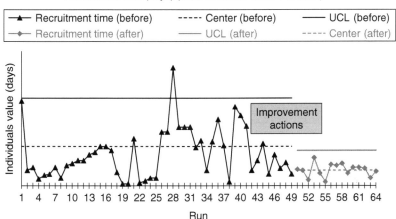

Individuals chart with Shewhart control limits

Recruitment time (days) (from–to before & from–to after)

Figure 13.6 Control chart of recruitment time in resourcing process.

Overall outcomes of improvement activities are HR processes with embedded continuous improvement that transforms problems into preventive actions. HR employees' effort and results were recognized, for example a number of verbal praises from internal customers, even small gifts, previously unheard of in HR function. It was acknowledged that HR measure themselves and are becoming proactive. The future of internal HR department started to look more secure. HR team enjoyed improvement journey, and themselves were perceived as moving from compliant to committed employees.

Intangible costs of poor quality have been recognized, for example throughput time in the Recruitment process was evaluated in £££'s per day delay. Six Sigma in HR processes has demonstrated values and benefits of proactive behavior, customer focus and alignment. Many skeptics have acknowledged HR Team improving their services. This experience enabled HR to assume the role of coach for the deployment of Six Sigma in the business.

13.9 Conclusion

Six Sigma in HR processes verified what the author has already experienced over a number of years of implementation of Six Sigma in different manufacturing and service processes across various organizations, as summarized in Table 13.4.

Six Sigma is not a panacea, it is a structured approach of using best practices to make right vision become reality. Leadership still needs to provide vision, drive and resources.

Figure 13.7 presents evaluation of the level of tools application, impact on results achieved and simplicity of the tools deployment (e.g., from 'team can just do it' to 'team needs a comprehensive training, facilitation and coaching,' relevant to this project).

Table 13.4 Six Sigma strengths, weaknesses, opportunities and threats (SWOT) analysis, based on Six Sigma implementation across many organizations and processes, including 'Six Sigma in HR' project

Strengths

1. Statistical thinking – facts-based decision-making, producing the new knowledge and knowledge-based management
2. Operational infrastructure, that is pre-defined roles and responsibilities of the Board, champion, mentors, Master Black Belts, Black Belts, Green Belts, Yellow Belts...
3. Effective, simple and universally applicable statistical tools
4. Supply chain process focus cuts across departments, functions, organizations, countries... functions, bringing in a 'common language'
5. Mechanics/process of projects, teamwork, improvements management
6. Focus on the right, customers' defined CTQ, Zb metrics

Weaknesses

1. When abusing statistics 'Business politicians' can alienate most of their organization with 'Six Sigma'
2. When prescribed Six Sigma infrastructure is imposed for the sake of it, with no consideration to a particular organization needs and goals
3. Paralysis by analysis, 'overkill' with analytical and statistical tools and more importantly presentation
4. The weakest link in a supply chain, who is strong in decision-making powers, might block progress (similar issue as in business process re-engineering)
5. Potential risk is when 'hard' issues (mechanics of Six Sigma) are given priority over 'soft' issues (e.g., communication, team building, motivation)
6. Pursuing achievement of 6 Zb for non-relevant metrics

Opportunities

1. When implemented as a business improvement strategy
2. Almost 20 years of Six Sigma good track record, proven across many industries and countries
3. Complements well with lean approach and tools
4. When projects' objectives and people are aligned (Six Sigma might not necessarily achieve this itself)

Threats

1. When implemented as a substitute to business strategy
2. Seeing Six Sigma failures as caused by the methodology itself, instead of understanding what other, non-Six Sigma factors, cause lack of progress
3. Lack of understanding might create an image of Six Sigma as a competitor to lean
4. Wrong projects (e.g., 'pet' projects) and/or misaligned people's objectives

A number of key success factors have been identified during this and other Six Sigma projects:

- 'Soft' (people) issues make 80% of success (80% is an estimated number, based on feedback from various improvement teams, and interviews with a number of Six Sigma practitioners, collected over a period of over 10 years).
- Achievement of goals and objectives starts by shared and clear vision.
- Teamwork, based on team building, training, coaching and delegating/ empowerment.
- Participative mentor's support (in most cases mentor's bi-weekly review is not good enough).
- Experienced facilitator/coach, who assures Six Sigma effectiveness by using right tools at right time and provides just-in-time training, training refreshers, facilitation or coaching.

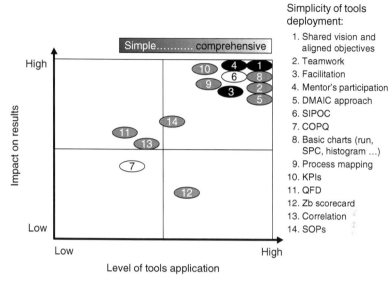

Figure 13.7 below refers to a chart:

Simplicity of tools deployment:

1. Shared vision and aligned objectives
2. Teamwork
3. Facilitation
4. Mentor's participation
5. DMAIC approach
6. SIPOC
7. COPQ
8. Basic charts (run, SPC, histogram ...)
9. Process mapping
10. KPIs
11. QFD
12. Zb scorecard
13. Correlation
14. SOPs

Figure 13.7 Experience of Six Sigma tools in the project.

- Involvement of internal/external suppliers and customers.
- Process mapping and SIPOC of 'as-is' and 'to-be.'
- Effective and efficient CHART data collection system.
- Use of simple charts (Pareto, histograms, bar-charts, run-charts ...).
- Importance of criteria for KPIs selection.

Some specifics to application of Six Sigma in HR processes have also been identified:

- Lower initial credibility of Six Sigma in HR; seen as 'statistics.'
- Harder definition of processes' scope and higher impact of perceptual elements.
- More of 'human factor' – direct dealing with people which usually turns into either a blocker and frustration or a driver and rewarding.
- High intangible cost of poor quality (e.g., communication).
- Less tangible measurements require more creative approach.
- Higher variety and lower predictability of individual customer's requirements.

SUCCESS

Success is achieved by
passionately sharing the vision
and motivating and leading the people to
complete right tasks in right way
that make the vision happen

References

Harrison, A. (1994). Six Sigma Training and Certification by Dr Mikel J. Harry. Private notes.

Harry, M.J. (1994). *The Vision of Six Sigma: A Roadmap for Breakthrough.* Phoenix, AZ: Six Sigma Publishing Company.

Huff, D. (2004). *How to Lie with Statistics.* Penguin Business.

Patterson, M. *et al.* (1997). *Impact of People Management Practices on Business Performance IPD,* London: The Cromwell Press, 10 yr study of 100 UK S/MB by LSE & IWP.

Wyper, B. and Harrison, A. (2000). Deployment of Six Sigma Methodology in Human Resource Function: A Case Study. *Total Quality Management,* 11(4 & 5), S720–S727.

www.onesixsigma.com

Scottish Engineering Lean Six Sigma Club: www.scottishengineering.org.uk

Index

For Product Safety Concerns and Information please contact our EU
representative GPSR@taylorandfrancis.com Taylor & Francis Verlag GmbH,
Kaufingerstraße 24, 80331 München, Germany

Printed and bound by CPI Group (UK) Ltd, Croydon, CR0 4YY
11/05/2025
01866585-0001